Cooperative Control and Optimization

Applied Optimization

Volume 66

Series Editors:

Panos M. Pardalos
University of Florida, U.S.A.

Donald Hearn
University of Florida, U.S.A.

The titles published in this series are listed at the end of this volume.

Cooperative Control and Optimization

Edited by

Robert Murphey

*Air Force Research Laboratory,
Eglin, Florida, U.S.A.*

and

Panos M. Pardalos

*University of Florida,
Gainesville, Florida, U.S.A.*

KLUWER ACADEMIC PUBLISHERS
DORDRECHT / BOSTON / LONDON

A C.I.P. Catalogue record for this book is available from the Library of Congress.

ISBN 978-1-4419-5217-2   e-ISBN 978-0-306-47536-8

Published by Kluwer Academic Publishers,
P.O. Box 17, 3300 AA Dordrecht, The Netherlands.

Sold and distributed in North, Central and South America
by Kluwer Academic Publishers,
101 Philip Drive, Norwell, MA 02061, U.S.A.

In all other countries, sold and distributed
by Kluwer Academic Publishers,
P.O. Box 322, 3300 AH Dordrecht, The Netherlands.

Printed on acid-free paper

" Things taken together are wholes and not wholes, something is being brought together and brought apart, which is in tune and out of tune; out of all things there comes a unity, and out of a unity all things."

- Heraclitus

> "

Contents

Contents

Preface

A cooperative system is defined to be multiple dynamic entities that share information or tasks to accomplish a common, though perhaps not singular, objective. Examples of cooperative control systems might include: robots operating within a manufacturing cell, unmanned aircraft in search and rescue operations or military surveillance and attack missions, arrays of micro satellites that form a distributed large aperture radar, employees operating within an organization, and software agents. The term entity is most often associated with vehicles capable of physical motion such as robots, automobiles, ships, and aircraft, but the definition extends to any entity concept that exhibits a time dependent behavior. Critical to cooperation is communication, which may be accomplished through active message passing or by passive observation. It is assumed that cooperation is being used to accomplish some common purpose that is greater than the purpose of each individual, but we recognize that the individual may have other objectives as well, perhaps due to being a member of other caucuses. This implies that cooperation may assume hierarchical forms as well. The decision-making processes (control) are typically thought to be distributed or decentralized to some degree. For if not, a cooperative system could always be modeled as a single entity. The level of cooperation may be indicated by the amount of information exchanged between entities. Cooperative systems may involve task sharing and can consist of heterogeneous entities. Mixed initiative systems are particularly interesting heterogeneous systems since they are composed of humans and machines. Finally, one is often interested in how cooperative systems perform under noisy or adversary conditions.

In December 2000, the Air Force Research Laboratory and the University of Florida College of Engineering successfully hosted the first Workshop on Cooperative Control and Optimization in Gainesville, Florida. About 40 individuals from government, industry, and academia attended and presented their views on cooperative control, what it means, and how it is distinct or related to other fields of research. This book contains se-

lected refereed papers summarizing the participants' research in control
and optimization of cooperative systems.

We would like to take the opportunity to thank the authors of the
papers, the Air Force Research Laboratory and the University of Florida
College of Engineering for financial support, the anonymous referees, S.
Butenko for preparing the camera ready manuscript, and Kluwer Academic Publishers for making the conference successful and the publication of this volume possible.

Robert Murphey and Panos M. Pardalos
August 2001

Chapter 1

COOPERATIVE CONTROL FOR TARGET CLASSIFICATION

P. R. Chandler
Flight Control Division
Air Force Research Laboratory (AFRL/VACA)
Wright-Patterson AFB, OH 45433-7531
phillip.chandler@va.afrl.af.mil

M. Pachter
Department of Electrical and Computer Engineering
Air Force Institute of Technology (AFIT/ENG)
Wright-Patterson AFB, OH 45433-7765
mpachter@afit.af.mil

Kendall E. Nygard
Department of Computer Science and Operations Research
North Dakota State University
Fargo, ND 58105-5164
nygard@cs.ndsu.edu

Dharba Swaroop
Department of Mechanical Engineering
Texas A & M University
College Station, TX 77842-3123
dswaroop@mengr.tamu.edu

Abstract   An overview is presented of ongoing work in cooperative control for unmanned air vehicles, specifically wide area search munitions, which perform search, target classification, attack, and damage assessment. The focus of this paper is the cooperative use of multiple vehicles to

R. Murphey and P.M. Pardalos (eds.), Cooperative Control and Optimization, 1–19.
© *2002 Kluwer Academic Publishers.*

maximize the probability of correct target classification. Capacitated transhipment and market based bidding are presented as two approaches to team and vehicle assigment for cooperative classification. Templates are developed and views are combined to maximize the probability of correct target classification over various aspect angles. Optimal trajectories are developed to view the targets. A false classification matrix is used to represent the probability of incorrectly classifying nontargets as targets. A hierarchical distributed decision system is presented that has three levels of decomposition: The top level performs task assignment using a market based bidding scheme; the middle subteam level coordinates cooperative tasks; and the lower level executes the elementary tasks, eg path planning. Simulations are performed for a team of eight air vehicles that show superior classification performance over that achievable when the vehicles operate independently.

Keywords: cooperative control, autonomous control

1.   Introduction

The wide area search weapon system, as presently envisioned [1] has a number of air vehicles operating independently. The vehicles are released in a target area, and follow a set of waypoints that are preset at launch. If an object is detected in the sensor footprint, the vehicle tries to classify the object as a target. If the classification satisfies the criteria, the vehicle attacks the target and is destroyed. To maximize the probability of finding high value targets in a short period of time, cooperation among the vehicles has been proposed and work is ongoing in developing cooperative control algorithms. Cooperative search algorithms are being pursued in [2] where a cognitive map of threats, targets, and terrain is constructed using sensor inputs from all the vehicles. Cooperative classification algorithms are being developed by the authors that combine aspect angle dependent views of an object from multiple vehicles to maximize the probability of correct classification. Cooperative attack algorithms are being developed in [3] to ensure that sufficient weapons are engaged to ensure destruction of the target. The weapon target assignment problem is being addressed in [4] using dynamic stochastic prgramming and in [5] using dynamic network flow optimization models. Online optimal trajectory generation for cooperative rendezvous has been pursued by the authors [6] and others [7, 8].

Cooperative classification as discussed in Section 2 is the task of optimally and jointly using multiple vehicles to maximize the probability of correct target classification. This is shown in Fig. 1.1 where the arrows represent the velocity vectors of the vehicles. The vehicles can communicate over some fixed range, which is represented by the large circles. The

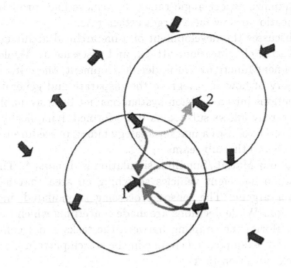

Figure 1.1. Target Classification Scenario

vehicle at 1 has detected a potential target, but in general the probability of classification $P_C < 1$ or some threshold. The vehicle could perform a loopback maneuver or another vehicle could view the potential target at a different aspect angle. An approach to combining the views statistically is given in the next section. An important issue is choosing the optimal aspect angle for the second view. Initially, the second view was chosen to be orthogonal to the first.

In Fig. 1.1 the vehicle at 2 also has detected a potential target. The vehicle at 3 could view either potential target 1 or 2, or the vehicle at 4 could view potential target 1. Determining which vehicle could optimally provide the second view is the assignment problem. The optimization function could be, among other alternatives, to maximize the value of targets classified, or to minimize the time to classify. Two approaches to the assignment problem are given in Section 3.

The mission performance of wide area search munitions is quite sensitive to false target attack rate. This stems from the sensor used, the capability of the sensor processing algorithm, the number and type of objects in the search area, and whether the objects are partially hidden in clutter. The basic approach is to observe the object at the optimum aspect angle, as discussed in Section 2, as well as over the largest range of aspect angles, at minimum cost. Cost is defined as detraction from search time or attack tasks. The cooperative classification uses adjacent

vehicles to maximize aspect angle ranges to achieve high probability of
correct classfiication or low false target attack rate.

Section 4 discusses the development of a hierarchical architecture for
cooperative search, classification, attack, and assessment. While many
of these component functions are under development, the critical organizational theory of how to integrate the disparate and generally contradictory functions into a decision system has not been available. The
three level hierarchy allows sub-teams to be formed dynamically at the
midlevel. The top level uses a market analogy biding procedure to assign
vehicles and tasks to the sub-teams.

In Section 5, our Matlab/Simulink simulation is discussed. This high
fidelity simulation has eight vehicles searching an area that has both
targets and nontargets. The sensor processing is emulated, including
false classification. While decisions are made concerning which task each
vehicle is to perform, the coupling between the tasks is not completely
accounted for. For example, the change in the search pattern if a vehicle
is assigned a classification task.

Section 6 discusses many of the issues in cooperative classification
that have yet to be addressed. The classification performance is also
discussed. Section 7 presents the conclusions.

2.   Joint classification

The key technique for achieving a low false target attack rate is to use
multiple views. A notional template is shown in Fig. 1.2 of probability
of correct classification versus the aspect angle (θ) at which the object
is viewed. This can also be looked at as a confidence level. False classification is addressed later in this section. To keep the occurance of false
classification low, the threshold is set high, in this case, $P_C > .9$ before
the target can be attacked. As can be seen in Fig. 1.2, the threshold is
achieved only over a narrow range of aspect angles (0 or 180 deg).

The objective is to combine the statistics from multiple views; in general, for two views:

$$P_C = P(\theta_1) + P(\theta_2) - P(\theta_1, \theta_2) \tag{1}$$

If the views are statistically independent:

$$P_C = P(\theta_1) + P(\theta_2) - P(\theta_1)P(\theta_2) \tag{2}$$

Initially it is assumed the views are uncorrelated. This assumption will
be relaxed later in the section. From eqn. 2, one can see that if the object
is viewed at the same θ, P_C increases. Intuitively, this is not reasonable,
since there is no additional information. It is generally true that if the

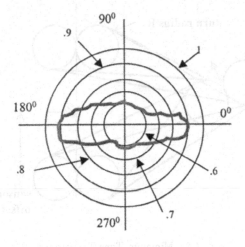

Figure 1.2.   Notional Target Template

aspect angles are separated by 90 deg the views should be uncorrelated and the information content should be greater. Based on this insight, trajectories now need to be derived that result in views orthogonal to the first.

Fig. 1.3 shows a sample configuration for two vehicles. The X marks the location of an object and the arrow the velocity vector of the vehicle that detected the object. The other arrow represents the velocity vector of an adjacent vehicle that could provide a second view. As stated earlier, the simplification is that the second vehicle should come in ±90 deg. Because the sensor looks ahead of the vehicle, the vehicle must be on the orthogonal line at least the distance of the sensor offset. The circles represent a specified minimum turn radius R.

It can be proven that the minimum time trajectory to a target consists of an initial turn through a circular arc, a straight line dash, and a turn through a final circular arc. The arcs are on the circles in Fig. 1.3. As can be seen, there are eight possible trajectories for the adjacent vehicle to place it's sensor on the object. The approach pursued here, is to calculate the distances traveled for all eight trajectories. The shortest, of course, is the minimum time trajectory. It can be shown that this algorithm holds for any configuration of adjacent vehicle and object. Once the trajectories are defined, we return to the target templates.

A notional target is shown in Fig. 1.4. For simplicity, the target is

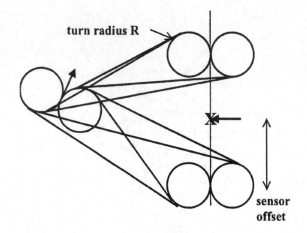

Figure 1.3.   Minimum Time Trajectories

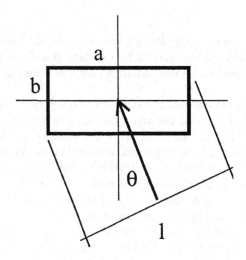

Figure 1.4.   Target Projection Diagram

rectangular with sides a, b and θ is the aspect angle at which the target is viewed. The assumption is that the projected line length l is proportional to the probability of classification. Views at $-\pi, -\pi/2, 0, \pi/2, \pi$

contain projections from only one side, so that no estimate of aspect ratio can be made. The probability, or confidence level, is defined as being proportional to the length of the side that is viewed. The projected line length is normalized by the length $a + b$, where the maximum occurs at θ^*. The orientation of the target on the ground is defined by ψ and θ. The probability (projection) is:

$$
P_c(\theta) = \begin{cases}
\frac{a\cos\theta + b\sin\theta}{a+b} & \text{for} \quad 0 \le \theta \le \pi/2 \\
\frac{-a\cos\theta + b\sin\theta}{a+b} & \text{for} \quad \pi/2 \le \theta \le \pi \\
\frac{-a\cos\theta - b\sin\theta}{a+b} & \text{for} \quad \pi \le \theta \le 3\pi/2 \\
\frac{a\cos\theta - b\sin\theta}{a+b} & \text{for} \quad 3\pi/2 \le \theta \le 2\pi
\end{cases}
$$

The maximum projected line length occurs at $\theta^* = \arctan(b/a)$. The maximum value is:

$$
P_c(\theta^*) = \frac{\sqrt{a^2 + b^2}}{a + b} \le 1
$$

Fig. 1.5 shows the periodic nature of $P_c(\theta)$, since a rectangle has 2 axes of symmetry. Finally, the plot is scaled by P_{max}, which in the figure is

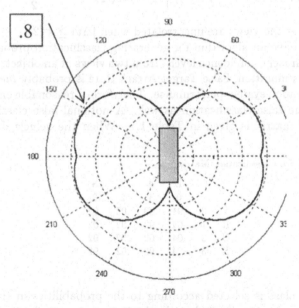

Figure 1.5. Probability versus aspect angle

.8. If the threshold is .9, this means that it is not possible to classify the target from one view.

If n statistically independent views of the target have been taken, the probability of identifying the target is calculated as:

$$P_{CI} = 1 - \Pi_{i=1}^{n}[1 - P_c(\theta_i)]$$

In the special case of $n = 2$, as before

$$P_{CI} = P_c(\theta_1) + P_c(\theta_2) - P_c(\theta_1)P_c(\theta_2)$$

The joint probability calculation given above is overly optimistic when the aspect angles are close. The exact joint probability for 2 views is not available, but it is reasonable that correlation $r = 1$ when $|\theta_2 - \theta_1| = \Delta\theta = 0$ and $r = 0$ when $\Delta\theta = 90°$. Therefore, as an approximation, a blending function is defined as:

$$\rho(\Delta\theta) = 1 - e^{-.03|\Delta\theta|}$$

The modification to the 2 view probability for correlated views is as follows:

$$P_{CC} = P_{CI}(\theta_1, \theta_2) + (1 - \rho(\Delta\theta))\left[P_c(\theta_1)P_c(\theta_2) - \frac{P_c(\theta_1) + P_c(\theta_2)}{2}\right]$$

This assumes the views are uncorrelated when $|\Delta\theta| \geq \pi/2$.

We now have an algorithm for generating classification probabilities, or more acurately, confidence levels, from two views of an object. For the autonomous munition, False Target Attack Rate is probably *the* critical factor in weapon system performance. Therefore, a reasonable emulation must include false classification as well. A notional false classification probability matrix is given in Table 1.1. When the vehicle detects a

Table 1.1. False Classification Matrix

		C	L	A	S	S
		0	1	2	3
T	0	.91	.03	.03	.03
R	1	.04	.92	.02	.02
U	2	.04	.02	.92	.02
E	3	.04	.02	.02	.92

target, the class is selected according to the probabilities in the table. The emulation enters the "selected" class template, not the "true" class template, at θ to get P_c. If $P_c < .9$, then another view is needed. If the second class is the same as the first, then proceed to combine views as above. If the combined statistic does not exceed the threshold, then the allocation process will determine if taking another view is cost effective. If the two classes are not the same, then the view with the highest priority class could be retained for the assignment and the other view discarded. An alternative is to retain both views and let the assignment algorithm determine from the priorities if an additional view is warrented. Cooperative target classification is driven by inputs from the upper level of the hierarchical cooperative control system currently under development. In this overview paper, we outline in the following 2 sections the assignment algorithms and hierarchical control architecture.

3.   Assignment

In general, the assignment problem involves not only classification, but also search, attack, and damage assessment. For purposes of illustration here, all of the vehicles are considered available to perform cooperative classification. The assignment algorithm then, is to select the optimal vehicle to provide the second view. An assignment method that includes the other tasks is addressed later in this section.

When an object is detected, the location, heading angle ψ, probability, and aspect angle θ is transmitted to all the other vehicles. The vehicles use Fig. 1.3 to determine distances and time to the object at, initially, angles perpendicular to ϕ. Later on, four angles to the object were used, these represent the best vectors to view the object from the template in Fig. 1.5. The calculated minimum time, distance, or cost to the object is then transmitted to the other vehicles. This is done for all the objects that need classification. The result is that all the vehicles have complete information and solution of the assignment problem is globally optimal. All the vehicles solve the same problem and therefore arrive at the same solution – conflicts are avoided and a degree of redundancy is achieved.

An example assignment matrix is given in Table 1.2. The columns

Table 1.2. Assignment Matrix

	T_1	T_2	T_3	T_4
V_1	C_1	C_2	C_3	C_4
V_2	C_5	C_6	C_7	C_8
V_3	C_9	C_{10}	C_{11}	C_{12}
V_4	C_{13}	C_{14}	C_{15}	C_{16}
V_5	C_{17}	C_{18}	C_{19}	C_{20}

are targets, the rows are vehicles, and the entries are costs, for example: time to object; remaining life; distance; or a weighted target value. Each of these types of costs have been used in the simulations discussed in a later section. This is a straightforward linear assignment problem and can be put in an integer linear programming form. This is easily solvable, even on modest hardware, for many targets and vehicles. The matrix is completely dynamic. As new objects are found, all of the vehicles are optimally reassigned. Or, when classified, taken out of the assignment matrix. The objective could also be to maximize the vehicles remaining life or to maximize the value of objects classified. The next topic addresses assignment for all the tasks.

The assignment of vehicles to search, classification, attack, and battle damage assessment is posed as a network flow optimization model. The model shown in Fig. 1.6 is described in terms of supplies and demands for a commodity, nodes which model transfer points, and arcs that interconnect the nodes and along which flow can take place. Arcs can have capacities that limit the flow along them. An optimal solution is the globally greatest benefit set of flows for which supplies flow through the network to meet the demands. In the model, vehicles are supplies and the tasks are demands. Since the vehicles are in only one mode at a time, the arcs have a flow of 0 or 1. Fig. 1.6 is also known as a Capacitated

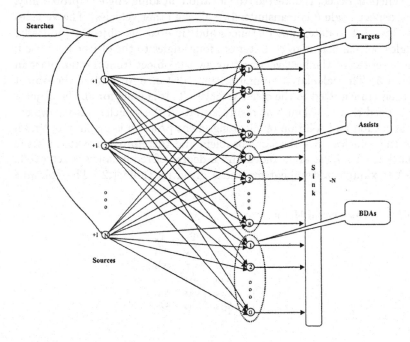

Figure 1.6. Network Flow Analogy

Network Trans-shipment Model and reduces to an integer (binary) linear programming problem. The linear program is formulated as follows:

$$Z = \max \sum_{i,j \epsilon I, i \neq j} c(i,j)x(i,j) \tag{3}$$

Subject to:

$$\sum_{j \epsilon I, i \neq j} x(i,j) - \sum_{k \epsilon I, k \neq i} x(k,i) = 0 \qquad i \epsilon I \tag{4}$$

$$\sum_{(i,j) \epsilon A} x(i,j) \leq b(i,j) \quad [(i,j)|i,j \epsilon I, i \neq j] \tag{5}$$

$$x(i,j) \geq 0 \quad [(i,j)|i,j \epsilon I, i \neq j] \tag{6}$$

where,
$c(i,j)$ = Expected value of vehicle i attacking target j,
$c(i,k)$ = Expected value of vehicle i classifying target k,
$c(i,g)$ = Expected value of vehicle i BDA target g,
$c(i,s)$ = Expected value of vehicle i continuing to search,
the vector x is the binary decision variable, c are the benefits to be maximized, Eqn. 4 is the flow balance, and Eqn. 5 is the link or flow capacity.

As in the previous assignment problem, the solution is globally optimal. The LP problem has a specialized structure that is very fast to solve, is highly flexible, event driven, and dynamic. An important issue is the determination of the costs above. Determining the utility of continuing to search may be particularly difficult to calculate. Also, output of the nodes are restricted to 1 to maintain linear form, which means multi-vehicle attack of a target is not allowed. This restriction is relaxed in the next section, either by augmenting the matrix or by a process of "bidding".

4.   Hierarchical architecture

Figure 1.7 illustrates a general architecture for cooperative control and resource allocation among multiple vehicles. The Inter-team Cooperative Planning agent is basically responsible for configuring teams of vehicles and providing them with their goals. This agent has visibility of the highest level goals for the overall mission, and it's internal model codifies doctrinal information as it applies at this level. If teams are preconfigured before takeoff, this agent will primarily be responsible for determining if teams should be reconfigured as new information is received and situation awareness improves. The models it invokes are expected to request and receive information from the Intra-team Cooperative Control Planning

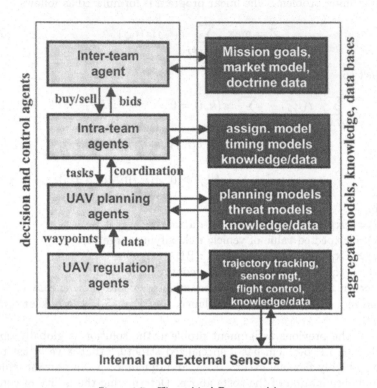

Figure 1.7. Hierarchical Decomposition

Agents at the next lower level. Based on this information, this agent may autonomously abandon certain high-level goals in favor of others.

The domain of responsibility for an Intra-team Cooperative Control Planning Agent involves the division of responsibilities among the vehicles working as a configured team. Leadership responsibiliites and coordination mechanisms depend on the mission, the models available to support accomplishing the mission, available data, and the current capabilites (eg fuel status) of the vehicles on the team.

The vehicle planning agents function specifically within the domain of an individual vehicle. These on-board planners accept a specific goal that is approprite for a single vehicle, then invoke path planning and scheduling algorithms aimed at meeting the goal.

Finally, the vehicle Regulating Agents provide command sequences for the vehicle, in order to accomplish such tasks as following trajectories, activating sensors, executing maneuvers, changing speed, and releasing weapons.

At the Inter-team level is a high-level auction procedure [10] for determining which targets should be assigned to which team. We assume that an initial allocation of targets to vehicles has been made, and that each team has solved its own generalized assignment problem to determine which vehicles attack which targets, and the total expected value of the chosen decisions. Thus, a team derives value through its current assets, which are targets to strike, and vehicles to strike them. To potentially improve the overall value among the teams, we now allow targets and vehicles to be "traded" from one team to another, in a way that simulates a stock exchange. When a team hypothetically gives up an asset, the following computations can be derived:

- For a specified target, the reduction in value to the team if the target is given to another team. This is the target "sell" value.

- For a specified vehicle, the reduction in value to the team if a vehicle is given to another team. This is the vehicle "sell" value.

Similarly, when a team acquires an asset, viz an additional target or vehicle, the following computations can be derived:

- The gain in value to the team if a specified target is received from another team. This is the target "buy" value.

- For a specified vehicle, the gain in value to the team if a vehicle is received from another team. This is the vehicle "buy" value.

The advantage of making a trade is guaranteed to be realized only if the trade is isolated from other trades involving the same teams, because the buy and sell values apply at the margin and the assigning of multiple vehicles to a target is inherently nonlinear.

The Intra-team level has agents that manage cooperative behavior, including: cooperative search, cooperative classification, cooperative attack, damage assessment, and rendezvous. Cooperative search consists of building maps of threats, targets, and terrain. As each of the vehicles uncovers information, it is transmitted to the other vehicles to build the maps. An optimization problem is solved to apportion individual vehicles to search areas that have the greatest probability of containing high value targets, while minimizing fuel and exposure. Cooperative classification has already been discussed. Cooperative attack stems from the probability of kill from an individual munition is less than one ($P_K \leq 1$). Multiple munitions may be needed to kill the target with sufficient confidence. Cooperative damage assessment is to ensure that high value targets have indeed been destroyed by viewing the target after attack. The rendezvous function is the time coordination of vehicles arrival at a target.

5.   Simulation

A simulation was developed for up to eight vehicles cooperatively controlled in a wide area search and attack mission. The simulation is based on the Control Automation and Task Allocation (CATA) [9] simulation in C++. The simulation was converted to run under Matlab Simulink to expedite algorithm research. Much of the software is compiled C++ code that is incorporated into Simulink blocks. The research algorithms are coded using graphics or math script.

The simulation scenario entails eight vehicles searching a battle space that has six targets of various values and up to five nontargets. The vehicles are initially in a echelon formation and following a serpentine path. As targets are detected, vehicles are dynamically assigned to perform classification and attack. The search could be dynamically changed as vehicles are assigned so as to cover the areas that have the highest probability of containing a high value target. In this simulation, if a vehicle is reassigned back to search, it returns to the original sepentine path. All of the targets are found and attacked before the vehicles run out of fuel. No nontargets are attacked.

Fig. 1.8 shows a typical scenario.

In Fig. 1.8 vehicle 2 detects target 2 first. Vehicle 5 is assigned to classify target 2. Vehicle 2 then detects target 6 and vehicle 3 detects target 5. Vehicle 6 is then assigned to classify target 5 and vehicle 7 is assigned to classify target 6. Then vehicle 3 detects target 7, which results in vehicle 8 being assigned to classify target 7. At this point, vehicle 2 detects target 4, which results in vehicle 4 being assigned to classify target 4. Vehicle 5 also detects target 5 on it's way to classify target 2, but this does not trigger a reassignment. This is because vehicle 5 does not pass over the target close enough to the specified aspect angle. The fortuitious detections could be more optimally incorporated. Finally, vehicle 3 detects target 3, however, vehicle 8 is assigned target 3. This results in vehicle 1 being assigned to target 7, where vehicle 1 had not been previously assigned. Vehicles 2 and 3 continue on the serpentine path, while the other vehicles classify and attack their assigned targets. All of the vehicles cross over their assigned targets at the specified aspect angles and the classification threshold is crossed. All of the targets are of a high value, so the targets are attacked as soon as they are classfied – there is no delayed attack.

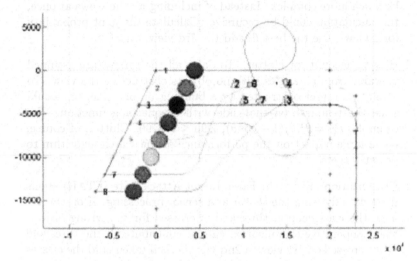

Figure 1.8. Scenario with 8 vehicles, 6 targets

Not shown, but if a false target attack rate and nontargets are introduced, all of the valid targets are attacked, but one of the nontargets is attacked. If the probability of correct classification threshold is raised, then the potential targets are viewed at more aspect angles. This prevents the nontargets from being attacked, but sometimes results in valid targets not being attacked. The developed simulation tool allows us to conduct a parametric study and thus optimally address this trade off situation.

6. Classification issues

1 Aspect angle estimate $\hat{\theta}$. This estimate could be used to determine the 2nd optimum viewing angle. To date, 2nd view angles are based on the heading angle ψ of the first view, not the orientation of the object on the ground. The computation of the optimal 2nd view could be done offline for the finite set of templates. However, this does not mean that the classification threshold will be crossed. Another offline optimization could be performed to determine the number and aspect angles of views to yield classification. Given that this information were available, the algorithms discussed previously could use it and allocate resources optimally.

2 Statistically combining 3 or more views. For a possible high value target, an arbitrary number of views may be desirable. The simplified joint probability approach presented earlier would have to be much more complex. Instead of including all the views at once, the calculation could be recursive. Calculate the joint probability for 2 views, use the best θ, add the 3rd view, etc.

3 How to account for clutter. To date, all the targets are assumed viewable from all angles with no objects obstructing the view. To emulate a target obstructed by a building on one side, one could scale the template on that side with a squashing function. For example, $P_C = P(\theta)(1 - \cos\theta)$, $-90 \leq \theta \leq 90$. Clutter, of course, has a large impact on the performance of the search algorithm to detect targets.

4 Classification mismatch. False Target Attack Rate (FTAR) stems most directly from the sensor and sensor processing. If on the 1st view the classification threshold is crossed for the wrong object, then cooperative classification cannot contribute. If the threshold is not crossed on 1st view, a 2nd view is then taken, and the classes are different, it is of course not possible to combine the statistics. However, the mismatch could be resolved by the optimization algorithm. Select one of the classes either based on probability of occurance or target value. The optimization then determines whether resources should be assigned to classify the selected object. If so, then the class from the 3rd view should break the tie. The additional views contribute to reducing the FTAR.

5 Other issues – registration. Previous discussions assume the multiple views are of the same object. This is a function of the navigation precison versus object density for fixed targets. If the objects can move between views, then registration is more of a concern and contributes to classification mismatch. Finally, if a vehicle is pulled off of search to perform a classification or other task, what is the impact on the search strategy? This is especially critical in mission performance if the objects to classify are ultimately nontargets.

7.   Conclusions

High fidelity simulations have been performed of eight vehicles in random and serpentine search patterns to detect, classify, and attack targets. Sensor processing is emulated using the target templates previously discussed. Work to date has focused on orthogonal 2nd views where the views are combined statistically. Minimum time maneuvers are used to view the potential target at the specified heading. The optimal assignment is based on minimizing the time to classify. Other metrics also used include: maximizing remaining life, and maximizing value of targets classified. The largest difference using these metrics is that maximizing remaining life resulted in delayed attacks until the vehicles were nearly out of fuel. Which of these functions are best in maximizing the probability of targets killed would come from a systems analysis study.

With communications and cooperative classification, fewer loop-back maneuvers are performed where the same vehicle performs the second view. This implies a more efficient utilization of resources. If the vehicles are in line formation where there is significant overlap in the sensor footprints, this results in extensive looping maneuvers to perform classification. Placing the vehicles in an echelon or staggered formation yields much more direct (efficient) classification trajectories.

The assignment techniques discussed are fast and globally optimal. The market approach to assignment becomes more useful as the number of vehicles increase; however, the benefit degrades as the transactions become more coupled.

Scenarios without coordination frequently result in valid targets not being found; cooperative classification successfully addresses this problem. Introduction of false classification can be countered with more emphasis on cooperative classification, but with some increase in the probability of not classifying valid targets.

Hierarchical cooperative control allows for near optimal solution of the large scale optimization problem. It is compatible with the prevailing information pattern in the air to ground attack acenario, and it is computationally efficient for dynamic replanning.

Acknowledgments

The authors wish to thank Lt. Col David Jacques and Dr. Robert Murphey for their contributions and support. The authors also wish to thank their colleague Steven Rasmussen for his technical support.

References

[1] Low Cost Autonomous Attack System briefing, approved for public release, AFDTC/PA, 24 Mar 99.

[2] Passino, Kevin M., Marios M. Polycarpu, David Jacques, and Meir Pachter, "Distributed Cooperation and Control for Autonomous Air Vehicles", ibid.

[3] Gillen, Daniel P., and David R. Jacques, "Cooperative Behavior Schemes for Improving the Effectiveness of Autonomous Wide Area Search Munitions", ibid.

[4] Murphy, Robert A., "An Approximate Algorithm for a Weapon Target Assignment Stochastic Program", in Approximation and Complexity in Numerical Optimization: Continuous and Discrete Problems, Klewer Academic Publishers, 1999.

[5] Nygard, Kendall E., Phillip R. Chandler, and Meir Pachter, "Dynamic Network Flow Optimization Models for Air Vehicle Resource Allocation", submitted to American Control Conference 2001.

[6] Chandler, P., S. Rasmussen, M. Pachter, "UAV Cooperative Control", AIAA GNC 2000, Denver, CO, Aug 2000.

[7] McLain, T., "Cooperative Rendezvous of Multiple Unmanned Air Vehicles", AIAA GNC 2000, Denver, CO, Aug 2000.

[8] Bortoff, S., "Path Planning for UAVs", AIAA GNC 2000, Denver, CO, Aug 2000.

[9] "Control Automation and Task Allocation", AFRL Final Report, Boeing, 1997.

[10] M. Wellman, and P. Wurman, "Market-aware Agents for a Multiagent World", Robotics and Autonomous Systems, 24:115-125, 1998.

Chapter 2

GUILLOTINE CUT IN APPROXIMATION ALGORITHMS

Xiuzhen Cheng, Ding-Zhu Du, Joon-Mo Kim and Hung Quang Ngo
Department of Computer Science and Engineering,
University of Minnesota, Minneapolis,
MN 55455, USA.
{cheng,dzd,jkim,hngo}@cs.umn.edu

Abstract   The guillotine cut is one of main techniques to design polynomial-time approximation schemes for geometric optimization problems. This article is a short survey on its history and current developments.

Keywords: approximation algorithms, guillotine cut

1.   Introduction

In 1996, Arora [1] published a surprising result that many geometric optimization problems, including the Euclidean TSP (traveling salesman problem), the Euclidean SMT (Steiner minimum tree), the rectilinear SMT, the degree-restricted-SMT, k-TSP, and k-SMT, have polynomial-time approximation schemes. More precisely, for any $\varepsilon > 0$, there exists an approximation algorithm for those problems, running in time $n^{O(1/\varepsilon)}$, which produces approximation solution within $1 + \varepsilon$ from optimal. It made Arora's research be reported in *New York Times* again. [1] Several weeks later, Mitchell [19] claimed that his earlier work [17] (its journal version [18]) already contains an approach which is able to lead to the similar results. However, one year later, Arora [2] made another big progress that he improved running time from $n^{O(1/\varepsilon)}$ to $n^3(\log n)^{O(1/\varepsilon)}$. His new polynomial-time approximation scheme also runs randomly in time $n(\log n)^{O(1/\varepsilon)}$. Soon later, Mitchell [20] claimed again that his approach can do a similar thing. We were curious about this piece of history and hence made a study on these two approaches. In this article,

21

R. Murphey and P.M. Pardalos (eds.), Cooperative Control and Optimization, 21–34.
© 2002 Kluwer Academic Publishers.

we would like to share with readers the result of our investigation and
something interesting that we found in their publications.

2.   Rectangular partition and guillotine cut

Let us start from rectangular partition. In fact, before prove his main
theorem, Mitchell [17, 18] stated clearly that "Our proof is inspired by
the proof in [7]" where the reference [7] in [17] ([9] in [18]) is actually
a paper of Du, Pan, and Shing [7] on minimum edge-length rectangular
partition. This paper initiated the idea of using guillotine cut to design
approximation algorithms.

The minimum edge-length rectangular partition (MELRP) was first
proposed by Lingas, Pinter, Rivest, and Shamir [13]. It can be stated
as follows: Given a rectilinear polygon possibly with some rectangular
holes, partition it into rectangles with minimum total edge-length.

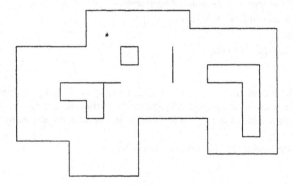

Figure 2.1.   Rectilinear polygon with holes.

The holes in the input rectangular polygon can be, possibly in part,
degenerated into a line segment or a point (Fig. 2.1).

There are several applications mentioned in [13] for the background of
the problem: "Process control (stock cutting), automatic layout systems
for integrated circuit (channel definition), and architecture (internal partitioning into offices). The *minimum edge-length* partition is a natural
goal for these problems since there is a certain amount of waste (e.g.
sawdust) or expense incurred (e.g. for dividing walls in the office) which
is proportional to the sum of edge lengths drawn. For VLSI design, this
criterion is used in the MIT 'PI' (Placement and Interconnect) System
to divide the routing region up into channels - we find that this produces
large 'natural-looking' channels with a minimum of channel-to-channel
interaction to consider."

They showed that the holes in the input make difference on the computational complexity. While the MELRP in general is NP-hard, the MELRP for hole-free inputs can be solved in time $O(n^4)$ where n is the number of vertices in the input rectilinear polygon. The polynomial algorithm is essentially a dynamic programming based on the following fact.

Through each vertex of the input rectilinear polygon, draw a vertical line and a horizontal line. Those lines will form a grid in the inside of the rectilinear polygon. Let us call this grid the *basic grid* for the rectilinear polygon (Fig. 2.2).

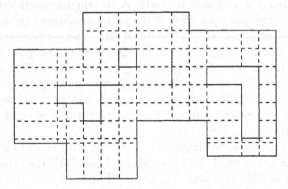

Figure 2.2. Basic grid.

Lemma 2.1 *There exists an optimal rectangular partition lying in the basic grid.*

Proof. Consider an optimal rectangular partition not lying in the basic grid. Then there is an edge not lying in the basic grid. Consider the maximal straight segment in the partition, containing the edge. Say, it is a vertical segment ab. Suppose there are r horizontal segments touching the interior of ab from right and l horizontal segments touching the interior of ab from left. If $r \geq l$, then we can move ab to the right without increasing the total length of the rectangular partition. Otherwise, we can move ab to the left. We must be able to move ab into the basic grid because, otherwise, ab would be moved to overlapping with another vertical segment, so that the total length of the rectangular partition is reduced, contradicting the optimality of the partition. □

A naive idea to design approximation algorithm for general case is to use a forest connecting all holes to the boundary and then to solve the resulting hole-free case in $O(n^4)$ time. With this idea, Lingas [14] gave the first constant-bounded approximation; its performance ratio is 41. Later, Du [9, 10] improved the algorithm and obtained a approximation with performance ratio 9. Meanwhile, Levcopoulos [15] provided a greedy-type faster approximation with performance ratio 29 and conjectured that his approximation may have performance ratio 4.5.

Motivated from a work of Du, Hwang, Shing, and Witbold [6] on application of dynamic programming to optimal routing trees, Du, Pan, and Shing [7] initiated an idea which is important not only to the MELRP problem, but also to many other geometric optimization problems. This idea is about guillotine cut. A cut is called a *guillotine cut* if it breaks a connected area into at least two parts. A rectangular partition is called a *guillotine rectangular partition* if it can be performed by a sequence of guillotine cuts. Du *et al* [7] noticed that there exists a minimum length guillotine rectangular partition lying in the basic grid, which can be computed by a dynamic programming in $O(n^5)$ time. Therefore, they suggested to use the minimum length guillotine rectangular partition to approximate the MELRP and tried to analyze the performance ratio. Unfortunately, they failed to get a constant ratio in general and only obtained a result in a special case.

In this special case, the input is a rectangle with some points inside. Those points are holes. It had been showed (see [11]) that the MELRP in this case is still NP-hard. Du *et al* [7] showed that the minimum length guillotine rectangular partition as approximation of the MELRP has performance rato at most 2 in this special case. The following is a simple version of their proof, published in [8].

Theorem 1 *The minimum length guillotine rectangular partition is a approximation with performance ratio 2 for the MELGP.*

Proof. Consider a rectangular partition P. Let $proj_x(P)$ denote the total length of segments on a horizontal line covered by vertical projection of the partition P.

A rectangular partition is said to be covered by a guillotine partition if each segment in the rectangular partition is covered by a guillotine cut of the latter. Let $guil(P)$ denote the minimum length of guillotine partition covering P and $length(P)$ the total length of rectangular partition P. We will prove

$$guil(P) \leq 2 \cdot length(P) - proj_x(P)$$

by induction on the number k of segments in P.

For $k = 1$, we have $guil(P) = length(P)$. If the segment is horizontal, then we have $proj_x(P) = length(P)$ and hence

$$guil(P) = 2 \cdot length(P) - proj_x(P).$$

If the segment is vertical, then $proj_x(P) = 0$ and hence

$$guil(P) < 2 \cdot length(P) - proj_x(P).$$

Now, we consider $k \geq 2$. Suppose that the initial rectangle has each vertical edge of length a and each horizontal edge of length b. Consider two cases:

Case 1. There exists a vertical segment s having length $\geq 0.5a$. Apply a guillotine cut along this segment s. Then the remainder of P is divided into two parts P_1 and P_2 which form rectangular partition of two resulting small rectangles, respectively. By induction hypothesis,

$$guil(P_i) \leq 2 \cdot length(P_i) - proj_x(P_i)$$

for $i = 1, 2$. Note that

$$
\begin{aligned}
guil(P) &\leq guil(P_1) + guil(P_2) + a, \\
length(P) &= length(P_1) + length(P_2) + length(s), \\
proj_x(P) &= proj_x(P_1) + proj_x(P_2).
\end{aligned}
$$

Therefore,

$$guil(P) \leq 2 \cdot length(P) - proj_x(P).$$

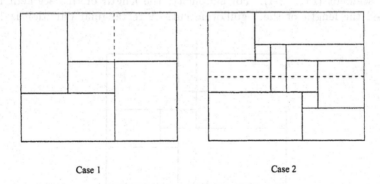

Case 1   Case 2

Figure 2.3. The proof of Theorem 1.

Case 2. No vertical segment in P has length $\geq 0.5a$. Choose a horizontal guillotine cut which partitions the rectangle into two equal parts. Let P_1 and P_2 denote rectangle partitions of the two parts, obtained from P. By induction hypothesis,

$$guil(P_i) \leq 2 \cdot length(P_i) - proj_x(P_i)$$

for $i = 1, 2$. Note that

$$
\begin{aligned}
guil(P) &= guil(P_1) + guil(P_2) + b, \\
length(P) &\geq length(P_1) + length(P_2), \\
proj_x(P) &= proj_x(P_1) = proj_x(P_2) = b.
\end{aligned}
$$

Therefore,

$$guil(P) \leq 2 \cdot length(P) - proj_x(P).$$

□

Gonzalez and Zheng [12] improved the constant 2 in Theorem 1 to 1.75 with a very complicated case-by-case analysis. Du, Hsu, and Xu [8] extended the idea of guillotine cuts to the convex partition problem.

3.   1-Guillotine cut

Mitchell [17, 18] gave an approximation with performance ratio 2 for the MELRP in the general case by extending the idea of guillotine cut.

First, he uses a rectangle to cover the input rectangular polygon with holes. Then, he extended the guillotine cut to the 1-guillotine cut. A *1-guillotine cut* is a partition of a rectangle into two rectangles such that the cut line intersects considered rectangular partition with at most one segment (Fig. 2.4). For simplicity, the length of this segment is called the length of the 1-guillotine cut. A rectangular partition is 1-

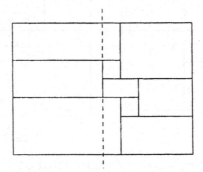

Figure 2.4.   1-guillotine cut.

guillotine if it can be realized by a sequence of 1-guillotine cuts (Fig. 2.5). The minimum 1-guillotine rectangular partition can also be computed by dynamic programming in $O(n^{16})$ time. In fact, at each step, the 1-guillotine cut has $O(n^4)$ choices. There are $O(n^4)$ possible rectangles appearing in the algorithm. Each rectangle has $O(n^8)$ possible boundary conditions.

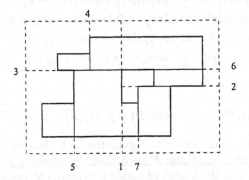

Figure 2.5. 1-guillotine rectangular partition with seven cuts.

To establish the performance ratio of the minimum 1-guillotine rectangular partition as an approximation of the MELRP, Mitchell [17] showed the following.

Theorem 2 *For any rectangular partition P, there exists a 1-guillotine rectangular partition P' covering P such that*

$$length(P') \leq 2length(P).$$

Proof. It can be proved by an argument similar to the proof of Theorem 1. Let $guil_1(P)$ denote the minimum length of a 1-guillotine rectangular partition covering P and $length(P)$ the length of the rectangular partition P. Let $proj_x(P)$ ($proj_y(P)$) denote the total length of segments on a horizontal (vertical) line covered by vertical (horizontal) projection of the partition P We will prove

$$guil_1(P) \leq 2 \cdot length(P) - proj_x(P) - proj_y(P)$$

by induction on the number k of segments in P.

For $k = 1$, we have $guil_1(P) = length(P)$. Without loss of generality, assume that the segment is horizontal. Then we have $proj_x(P) = length(P)$ and $proj_y(P) = 0$. Hence

$$guil_1(P) = 2 \cdot length(P) - proj_x(P) - proj_y(P).$$

Now, we consider $k \geq 2$ in the following two cases:

Case 1. There exists a 1-guillotine cut. Without loss of generality, assume this 1-guillotine cut is vertical with length a. Suppose the remainder of P is divided into two parts P_1 and P_2. By induction hypothesis,

$$guil_1(P_i) \leq 2 \cdot length(P_i) - proj_x(P_i) - proj_y(P_i)$$

for $i = 1, 2$. Note that

$$
\begin{aligned}
guil_1(P) &\leq guil_1(P_1) + guil_1(P_2) + a, \\
length(P) &= length(P_1) + length(P_2) + a, \\
proj_x(P) &= proj_x(P_1) + proj_x(P_2) \\
proj_y(P) &\leq proj_y(P_1) + proj_y(P_2).
\end{aligned}
$$

Therefore,

$$
guil_1(P) \leq 2 \cdot length(P) - proj_x(P) - proj_y(P).
$$

Case 2. There does not exist 1-guillotine cut. In this case, we need to add a segment to partition P such that the resulting partition has a 1-guillotine cut and the length of added segment is at most $proj_x(P_1) + proj_x(P_2) - proj_x(P)$ if the 1-guillotine cut is horizontal and at most $proj_y(P_1) + proj_y(P_2) - proj_y(P)$ if the 1-guillotine cut is vertical, where P_1 and P_2 are partitions obtained from P by the 1-guillotine cut. To do so, it suffices to show that there exists a line such that the length of added segment for the line to become a 1-guillotine cut is not more than the total length of segments on the line, receiving projection from both sides. For simplicity of description, let us call by *horizontal (vertical) 1-dark point* a point receiving horizontal (vertical) projection from both sides. Then, for a horizontal (vertical) line, the set of vertical (horizontal) 1-dark points form the segment adding which would make the line become a 1-guillotine cut.

Lemma 3.1 *Let H (V) be the set of all horizontal (vertical) 1-dark points. Then there exists either a horizontal line L such that*

$$
length(L \cap H) \leq length(L \cap V)
$$

or a vertical line L such that

$$
length(L \cap H) \geq length(L \cap V).
$$

Proof. First, assume that the area of H is not smaller than the area of V. Denote $L_a = \{(x, y) \mid x = a\}$. Then areas of H and V can be represented by

$$
\int_{-\infty}^{+\infty} length(L_a \cap H) da
$$

and

$$
\int_{-\infty}^{+\infty} length(L_a \cap V) da,
$$

respectively. Since

$$\int_{-\infty}^{+\infty} length(L_a \cap H)da \geq \int_{-\infty}^{+\infty} length(L_a \cap V)da,$$

there must exist a such that

$$length(L_a \cap H) \geq length(L_a \cap V).$$

Similarly, if the area of H is smaller than the area of V, then there exists a horizontal line L such that

$$length(L \cap H) \leq length(L \cap V).$$

$\square$

By Lemma 3.1, without loss of generality, we may assume that there exists a horizontal line L such that

$$length(L \cap H) \leq length(L \cap V),$$

that is,

$$length(L \cap H) \leq proj_x(P_1) + proj_x(P_2) - proj_x(P)$$

where P_1 and P_2 are subpartitions obtained from P by the line which becomes a 1-guillotine cut after adding segment $L \cap H$ to the partition P. By induction hypothesis,

$$guil(P_i) \leq 2 \cdot length(P_i) - proj_x(P_i) - proj_y(P_i)$$

for $i = 1, 2$. Note that

$$proj_y(P) \leq proj_y(P_1) + proj_y(P_2).$$

Therefore,

$$\begin{aligned} guil(P) &= guil(P_1) + guil(P_2) + length(L \cap H), \\ &\leq 2 \sum_{i=1}^{2} length(P_i) - \sum_{i=1}^{2} proj_x(P_i) - \sum_{i=1}^{2} proj_y(P_i) + length(L \cap H) \\ &\leq 2 \cdot length(P) - proj_x(P) - proj_y(P). \end{aligned}$$

$\square$

Mitchell [17, 18] used a different way to present the proof of Theorem 2. He symmetrically charged a half of the length of added segment to those parts of segments in P which face to 1-dark points. Since charge must be performed symmetrically, each point in P can be charged at most twice during the entire modification from a rectangular partition to a 1-guillotine rectangular partition. Therefore, the total length of added segments is at most $length(P)$ and hence Theorem 2 holds. Actually, this argument is equivalent to the current proof of Theorem 2. In fact, only projections from both sides exist (Case 2), $proj_x(P)$ or $proj_y(P)$ can contribute something against the length of the added segment.

4.   *m*-Guillotine cut

Mitchell [19] extended the 1-guillotine cut to the m-guillotine cut in the following way: A point p is a horizontal (vertical) m-*dark* point if the horizontal (vertical) line passing through p intersects at least $2m$ vertical (horizontal) segments of the considered rectangular partition P, among which at least m are on the left of p (above p) and at least m are on the right of p (below p). Let H_m (V_m) denote the set of all horizontal (vertical) m-dark points. An m-*guillotine cut* is either a horizontal line L satisfying

$$L \cap H_m \subseteq L \cap P$$

or a vertical line L satisfying

$$L \cap V_m \subseteq L \cap P.$$

A rectangular partition is m-guillotine if it can be realized by a sequence of m-guillotine cuts. The minimum m-guillotine rectangular partition can also be computed by dynamic programming in $O(n^{10m+6})$ time. In fact, at each step, an m-guillotine cut has at most $O(n^{2(m+1)})$ choices. There are $O(n^4)$ possible rectangles appearing in the algorithm. Each rectangle has $O(n^{8m})$ possible boundary conditions. By a similar argument, Mitchell [19] established the following result.

Theorem 3 *For any rectangular partition P, there exists an m-guillotine rectangular partition P' covering P such that*

$$length(P') \leq (1 + \frac{1}{m})length(P).$$

Corollary 4.1 *There exists a $(1 + \varepsilon)$-approximation with running time $n^{O(\log 1/\varepsilon)}$ for MELRP.*

From the 1-guillotine cut to the m-guillotine cut, it has no technical difficulty. But, why Mitchell did not do such an extention until Arara [1] published his remarkable results we mentioned at the beginning of this article? The answer is that before Arora's breakthrough, nobody was thinking in this way. Indeed, the importance of Arora's work [1] is more on opening people's mind than proposing new techniques. In terms of techniques, m-guillotine is more powerful. For example, we do not know how to apply Arora's techniques in [1] to obtain polynomial-time approximation schemes for the MELRP, the rectilinear Steiner arborescence problem [16], and the symmetric Steiner arborescence problem [5]. But, the m-guillotine cut works for them. For problems in high dimensional space, in particular, for geometric optimization problems in three or more dimensional space, Arora [1] provided $(1+\varepsilon)$-approximation with running time $n^{O(\log 1/\varepsilon (\log n)^{d-1})}$. But, the m-guillotine cut can still provide with polynomial-time approximation schemes. We will give more explanation in the next section.

5.   Portals

Arora's polynomial-time approximation scheme in [1] is also based on a sequence of cuts on rectangles. For example, let us consider Euclidean SMT. Initially, use a square to cover n input points. Then with a tree structure, partition this square into small rectangles each of which contains one given point. By choosing cut line in a range between 1/3 and 2/3 of an edge, Arora managed the tree structure to have depth $O(\log n)$.

To reduce the number of crosspoints at each cut line, Arora [1] use a different technique. This technique is the portal. The portals are points on cut line equally dividing cut segments. For Euclidean SMT (or Euclidean TSP, etc), crosspoints of the Steiner tree on a cut line can be moved to portals. This would reduce the number of crosspoints on the cut line. Suppose the number of portals is p. It can be proved that by properly choosing cut line, at each level of the tree structure moving crosspoints to portals would increase the length of the tour within three pth of the total length of the Steiner tree. Since the tree structure has depth $O(\log n)$, the total length of the resulting Steiner tree is within $(1+\frac{3}{p})^{O(\log n)}$ times the length of the optimal one. To obtain $(1+\frac{3}{p})^{O(\log n)} \leq 1 + \varepsilon$, we have to choose $p = O(\frac{\log n}{\varepsilon})$.

For problems in 3 or higher-dimensional space, the cut line should be replaced by cut plane or hyperplane. The number of portals would be $O((\frac{\log n}{\varepsilon})^2)$ or more. With so many possible crosspoints, the dynamic programming cannot run in polynomial time. However, the m-guillotine cut has at most $2m$ crosspoints in each dimension and m is a constant with respect to n. Therefore, the polynomial-time for the dynamic programming would be preserved under increasing dimension.

Combining the two techniques (the portal and the m-guillotine cut) can reduce the running time for dynamic programming. In fact, the portal technique first reduces the number of possible positions for crosspoints to $O(\frac{\log n}{\varepsilon})$ and this enables us to choose $2m$ from the $O(\frac{\log n}{\varepsilon})$ positions to form a m-guillotine cut $(m = 1/\varepsilon)$. Therefore, the dynamic programming for finding the best such partition runs in time $n^c(\log n)^{O(1/\varepsilon)}$ where c is a constant. This is the basic idea of Arora [2] and Mitchell [20]. Arora's work [2, 3] also contains a new technique about the tree structure of partition. Indeed, it is an earlier and better work compared with Mitchell [20].

The portal technique cannot apply to the MELRP, the rectilinear Steiner arborescence, and the symmetric rectilinear Steiner arborescence. In fact, for these three problems, moving crosspoints to portals is sometimes impossible. Therefore, it is an open problem whether there exists a polynomial-time approximation scheme with running time $O(n^c(\log n)^{O(1/\varepsilon)})$.

The power of the m-guillotine cut also has certain limitation. For example, we do not know how to establish a polynomial-time approximation scheme without including total length of given segments in the problem of interconnecting highways [4]. This provides some opportunities for further developments of those elegant techniques.

References

[1] S. Arora, Polynomial-time approximation schemes for Euclidean TSP and other geometric problems, *Proc. 37th IEEE Symp. on Foundations of Computer Science*, (1996) pp. 2-12.

[2] S. Arora, Nearly linear time approximation schemes for Euclidean TSP and other geometric problems, I*Proc. 38th IEEE Symp. on Foundations of Computer Science*, (1997) pp. 554-563.

[3] S. Arora, Polynomial-time approximation schemes for Euclidean TSP and other geometric problems, *Journal of the ACM* 45 (1998) 753-782.

[4] X. Cheng, J.-M. Kim and B. Lu, A polynomial time approximation scheme for the problem of interconnecting highways, to appear in *Journal of Combinatorial Optimization*.

[5] X. Cheng, B. DasGupta, and B. Lu, A polynomial time approximation scheme for the symmetric rectilinear Steiner arborescence problem, to appear in *Journal of Global Optimization*.

[6] D.Z. Du, F.K. Hwang, M.T. Shing and T. Witbold: Optimal routing trees, *IEEE Transactions on Circuits* 35 (1988) 1335-1337.

[7] D.-Z. Du, L.-Q. Pan, and M.-T. Shing, Minimum edge length guillotine rectangular partition, Technical Report 0241886, Math. Sci. Res. Inst., Univ. California, Berkeley, 1986.

[8] D.-Z. Du, D.F. Hsu, and K.-J Xu, Bounds on guillotine ratio, *Congressus Numerantium* 58 (1987) 313-318.

[9] D.-Z. Du, On heuristics for minimum length rectangular partitions, Technical Report, Math. Sci. Res. Inst., Univ. California, Berkeley, 1986.

[10] D.-Z. Du and Y.-J. Zhang: On heuristics for minimum length rectilinear partitions, *Algorithmica*, 5 (1990) 111-128.

[11] T. Gonzalez and S.Q. Zheng, Bounds for partitioning rectilinear polygons, *Proc. 1st Symp. on Computational Geometry*, 1985.

[12] T. Gonzalez and S.Q. Zheng, Improved bounds for rectangular and guillotine partitions, *Journal of Symbolic Computation* 7 (1989) 591-610.

[13] A. Lingas, R. Y. Pinter, R. L. Rivest, and A. Shamir, Minimum edge length partitioning of rectilinear polygons, *Proc. 20th Allerton Conf. on Comm. Control and Compt.*, Illinos, 1982.

[14] A. Lingas, Heuristics for minimum edge length rectangular partitions of rectilinear figures, *Proc. 6th GI-Conference*, Dortmund, January 1983 (Springer-Verlag).

[15] C. Levcopoulos, Fast heuristics for minimum length rectangular partitions of polygons, *Proc 2nd Symp. on Computational Geometry*, 1986.

[16] B. Lu and L. Ruan, Polynomial time approximation scheme for the rectilinear Steiner arborescence problem, *Journal of Combinatorial Optimization* 4 (2000) 357-363.

[17] J.S.B. Mitchell, Guillotine subdivisions approximate polygonal subdivisions: A simple new method for the geometric k-MST problem. *Proc. 7th ACM-SIAM Symposium on Discrete Algorithms*, (1996) pp. 402-408.

[18] J.S.B. Mitchell, A. Blum, P. Chalasani, S. Vempala, A constant-factor approximation algorithm for the geometric k-MST problem in the plane *SIAM J. Comput.* 28 (1999), no. 3, 771-781.

[19] J.S.B. Mitchell, Guillotine subdivisions approximate polygonal subdivisions: Part II - A simple polynomial-time approximation scheme for geometric k-MST, TSP, and related problem, *SIAM J. Comput.* 29 (1999), no. 2, 515-544.

[20] J.S.B. Mitchell, Guillotine subdivisions approximate polygonal subdivisions: Part III - Faster polynomial-time approximation scheme for geometric network optimization, preprint, 1997.

Chapter 3

UNMANNED AERIAL VEHICLES: AUTONOMOUS CONTROL CHALLENGES, A RESEARCHER'S PERSPECTIVE

Bruce T. Clough

Air Force Research Laboratory
Control Sciences Division
Wright-Patterson AFB, OH
Bruce.Clough@wpafb.af.mil

Abstract   AFRL is pressing ahead with development of truly autonomous UAV control systems. As we go from systems where the human is the pilot, through systems where the human is the operator, to systems where the human is the supervisor; with the ultimate goal simply to have the human as customer of UAV ops, we are running into numerous challenges. Yes, we face the typical technological questions of "What types of human tasks can we replace with on-board algorithms?" and "How big of a processor is required on-board to do this?". What are usually not asked are other questions, maybe not technically exciting, but with enormous practical impact: "How can we affordably add more code to already costly flight critical s oftware programs?" "How do I flight certify a system that has non-deterministic attributes?" "What is the impact of implementing distributed, coordinated, info-centric control systems that now have flight critical data links susceptible to electronic and information warfare?" "How do I convince the FAA, and foreign governments, that it's safe to let autonomous vehicles roam the skies?" These, and other questions, have just as great, if not greater, impact on systems development as the raw autonomous technology itself. This paper examines some of these challenges, how current AFRL research is addressing them, and points the way to future research that will allow truly autonomous operations.

Keywords: unmanned vehicles, autonomous control

R. Murphey and P.M. Pardalos (eds.), Cooperative Control and Optimization, 35–53.

1.   Introduction

Autonomous UAVs, these conjure up a host of visions in the minds of various researchers and developers. If one were to ask us what our vision of autonomous UAVs would be, we would refer them to the "flying monkeys" scene of the Wizard of Oz. Those "autonomous UAVs" exhibit behaviors that we would like to instill in our UAV development:

- They are self-organizing, taking high level goals "Get the Ruby Slippers" and translating them into the tasks required "fly there, find prey, capture, etc." without explicit instructions from the user.

- They are self-executing, not relying on external input to initiate behaviors.

- They are self-deconflicting, not relying on any external communications to keep from running into each other.

- They show battlefield management, are aware of targets and the threats in the terminal area, and take appropriate team action to achieve goals (scatter the Strawman, steal the Tin Man's axe, etc.).

- Oh, and they swarm!

ll kidding aside, our goal is to enable autonomous UAV operations for any USAF mission. Whether or not the missions are executed that way will be up to the policy makers, but they will have the technology needed if chosen. This is the view of autonomy development from a "shop" that accomplishes the transition of control technologies from basic research to technology demonstrations, in Department of Defense "lingo" from 6.1 to 6.3.

2.   Background

Before I press on to the challenges we face in enabling autonomous UAVs, I'd first like to cover a couple of definitions:

What are unmanned aerial vehicles?

The definition of "unmanned" simply means that a human is not aboard actively piloting or directing the aircraft. It might be carrying human cargo, but the control functions are either indigenous (on-board computer), or off-board (computer or remote pilot). We used to (up to about a year ago), use the term "uninhabited" rather than unmanned to be politically correct. We didn't want the audience to think that a human wasn't somewhere in the command & control process. Thankfully we've given that up!

What Is autonomy?

Autonomy, in our simple definition, is the capability of making human-type decisions without human aid. Simple task? No. Researchers have been working the autonomous systems area since the dawn of digital computers and we are just to the point where this research is starting to pay off. Autonomous vehicles of all types are being developed and used, from underwater exploration to the surface of Mars.

This said, what we are aiming at is not true autonomy, since humans will be somehow involved in autonomous system, at least in the systems the USAF is interested in. Our goal is to push this human interaction to the highest level possible.

Why do we want autonomous UAVs?

Now integrate the two definitions above. Why do this? Two main reasons; the first reason is cost.

- Eliminating on-board pilots eliminates quite a bit of vehicle size and complexity.

- If we build enough autonomy to turn the pilot into an operator (the continuous job of keeping the pointy-end forward is delegated to the on-board computer) then we eliminate a rated slot. This reduces the need to train pilots, reducing operation and support costs (O & S).

- Once an autonomous UAV is programmed, it won't change, so proficiency training is not required. The UAVs become wooden-rounds, kept in storage until needed and slashing O & S costs. Any training of the algorithms can be done with simulations.

The second reason is safety [Note 1]. This covers the dull, dangerous and dirty missions normally talked about as being the UAVs' missions.

- Dull - UAVs don't suffer fatigue, therefore they are well suited for missions that would bore human beings, such as constant ISR operations, or patrolling no-fly zones day-in and day-out.

- Dangerous - Being a military organization means something special to UAV operations, the knowledge that if the UAV doesn't return we don't have to mount a rescue mission, suffer through viewing a POW on CNN, or watch families grieve. We may have lost an asset, but no lives were lost. For instance, in Kosovo we lost 15+ UAVs, something that never made the front page of any news organ. Having UAVs gives us the option of sending vehicles into areas where we would be reluctant to send manned planes, giving field commanders new operational flexibility.

- Dirty - Searching out, and destroying, weapons of mass destruction (WMD) exposes the action agent to those same environments. UAVs are not biological systems, therefore they are immune to bio, chemical, and radiation effects, making them excellent scouts to find WMD, and in verifying damage after WMD neutralization strikes.

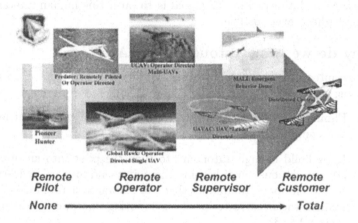

Figure 3.1.   The Autonomy Continuum

So where are we in developing autonomous vehicles?

Figure 3.1 is what we refer to as the Autonomy Continuum, with remotely piloted vehicles on the left to fully autonomous vehicles on the right.

Now in the field are remotely piloted vehicles such as Pioneer, Hunter, and Predator. Where we are going in the near term are remote operator vehicles such as Global Hawk and USAF UCAV. Here an operator tells the vehicle where to go and what to do, and the UAV responds. The operator is not a pilot, all pilot "point the nose" functions are accomplished by the vehicle.

Our current research is in the remote supervisor area, where an operator controls groups of UAVs. The UAVs are given a task to accomplish and "discuss" amongst themselves the local situation and how best to deal with it. We still see significant human involvement in some decisions, especially weapons targeting and release.

In the future the human becomes the customer to UAV capabilities, with the UAVs receiving high level goal oriented tasking from humans. They then determine the tasks, split up the "labor" and accomplish the mission.

3.   The Challenges

Basically, we can put the autonomous control challenges faced by Air Force researchers into three buckets: building intelligence, instilling safety, and enabling affordability.

Building intelligent, the stuff of everyday research

Building intelligence is the obvious challenge. How can we take the pilot out of the cockpit but leave the pilot functionality in the vehicle, especially the capability for handling unplanned situations and making life & death decisions? The pilot does a lot of things for an aircraft besides point the nose. Not only does the pilot manage the vehicles flight path, but also the pilot accomplishes vehicle health management, initiates and receives communications, does package management (if the pilot is the flight leader), and has to perform constant situational assessments (and re-assessment) to know what's going on. All these functions must be done onboard an autonomous air vehicle. Figure 3.2 shows this situation where the control system designer must take a holistic approach to integrating pilot functions into the autonomous software. Since all functions are integrated in the human, so must they be in the software.

Building intelligence is where most of the current UAV autonomous control research is focusing. This is the obvious and glamorous research area, but we feel that this cannot be the only research area. It could enable intelligence, but won't fly due to safety concerns and cost constraints. The three are intertwined - cost and safety considerations will influence how we build intelligence, not the other way around.

Instilling safety: If it isn't safe, it won't fly

The next challenge usually isn't considered by autonomous robotics researchers, but is at the top of the minds of most system implementers - safety. If the UAVs are going to be accepted as equal partners to manned aircraft, they must exhibit safety as good, or better than, manned aircraft [Ref 1].

The failure record of UAVs compared to manned aircraft is miserable. UAVs were considered toys by manned aircraft operators, and were con-

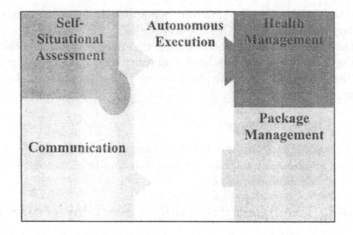

Figure 3.2. Replacing the Human: Holistic Approach

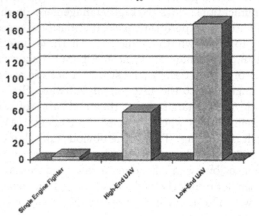

Figure 3.3. Comparison of Aircraft Loss Rates

sidered very unreliable. That's for good reason since the UAV mishap rate is a couple of magnitudes above manned aircraft. Figure 3.3 gives the LOA per 100,000 flight hours for three types of aircraft. "High-End UAVs" relate to the expensive UAVs such as Predator while the "Low-End UAVs" relate to systems such as Pioneer and Hunter. The manned data is from an aggregate of modern, single engine, fighter aircraft losses and shows an LOA of approximately 5E-5/flt-hr. So, by examining the data, the safety personnel's' concerns are borne out. Even the best UAVs are at least 10 times more prone to crash than modern single engine fighters.

This is not surprising when one considers that the usual system reliability spec is 80-90 % mission reliability. UAVs are inexpensive systems, their construction shows it, and the record proves it.

Our vision is to have UAVs capable of flying with manned aircraft in the same air space, possibly in the same formation. If mixed-fleet (manned & unmanned) operations are to be a reality, the reliability of the elements in the combined package must be comparable. In layman's terms, a human pilot wants to know that the unmanned "gizmo" flying with him/her is as reliable as he/she is.

The reliability and safety of manned aircraft flight control systems are linked through what we call PLOC (probability of loss of control) measured in system failures per flight hour [Note 2]. Manned aircraft PLOCs are 1E-6/flt-hr or less. Commercial specs are 1E-9. For systems to meet this, hardware and software go through rigorous and expensive development and testing procedures, and this increases cost, which brings us to the third challenge:

Enabling affordability: If we can't afford it we won't build it

Affordability is our final big challenge.

- We want systems to do the job that people did. This implies that the amount of software inside the flight control system must grow to accomplish the mission. Some estimates [Note 3] are for well over a million lines of higher-order code, an order of magnitude from where fielded systems are now.

- We want the systems to be safe as manned aircraft. This implies that the system must be tested as rigorous as manned systems, and many of the systems must exhibit good failure management so equipment failures do not result in unsafe conditions.

In other words, to be as safe (and as reliable) as a manned aircraft we must use manned-aircraft practices [Ref 2]. This means that the cost per vehicle of the UAV control systems will actually go up due to the added autonomy software and the fact that we can't eliminate existing parts of the systems due to reliability concerns [Note 2].

One could make the argument that this doesn't matter, since the increased production and procurement costs per capita will be more than offset by the O & S cost reductions, except we don't buy our systems based on O & S, we buy based on URF (unit reoccurring Flyaway) costs. This is driven by political considerations beyond our control, so we need to strive to reduce control system development costs as much as possible, and for autonomous control systems that means software costs.

Practicality constraints are # 1 in developmental engineering

As I stated in the Introduction, we transition technology into demonstrations, including flight demos. Therefore the technology not only has to work, but it must be safe and affordable. In fact, if one would rank these from a 6.3 flight test viewpoint, safety comes in first (if it's not safe, then it's a non-starter), followed by affordability (it must fit within the funding profile available), and then followed by performance (intelligence). The practical issues come to the forefront when implementing technology, so what course can we pursue to obtain safe, affordable, autonomous UAVs? The following are my thoughts on where we press ahead.

4.   How are we approaching these challenges?

Rules Of practical control systems

From years of developing flight control system software we have developed a few practical design rules which I would like to share. Where possible:

- Eliminate Feedbacks - Eliminating these eliminates sensors and/or lines of code required estimating values. There are fewer items to fail and less code to test.

- Eliminate Integrations - Integrations introduce time delays that add up to bust timing margins and require expensive code re-writes.

- Eliminate Communication - The fewer things have to communicate, the less chance there is for a foul-up. Communicate only those items that are critical [Note 6].

- Eliminate Explicit Models - Explicit models tend to be large, involving numerous integrations, and lack robustness to off-design conditions. Use implicit modeling wherever possible.

- Eliminate Code - Lines of code can be traded off for development cost. Current OFPs (the operation flight programs inside modern manned aircraft flight control computers) cost $ 500 per line to develop. A million lines of code demonstrating the reliability of manned systems could be cost prohibitive.

- Design for Good Enough, Not Optimal -Fielded systems fielded are not optimal, they are good enough. When we strive for optimal we always come up short due to us not knowing all the requirements. These same rules pertain to autonomous systems.

Our goal is to apply emerging technologies to develop practical autonomous systems, and for us that mean reducing cost and complexity where possible. In other words, promote simplicity. Unfortunately many of the traditional artificial intelligence techniques are machine hungry, relying on complicated algorithms to develop detailed explicit internal models that have to be run on state-of-the-art processors. In addition, many multi-UAV development efforts assume that one has infinite data availability with no latency or noise. These are recipes for practical system failure. We think the keys to practical operation are found in those studying natural systems with a nod to science fiction writers.

Technologies we are investigating

Biomimetics Simply put, biomimetics is development of novel materials, processes, and sensors through advanced understanding and exploitation of design principles found in natural biological systems. Translation: *Study how nature does something that you'd like to do and do it that way.* Nature always minimizes complexity to maximize survival; thus systems based on natural systems should mirror this simplicity. Two areas we are investigating closely are sensors and communications.

- Sensors are critical for autonomous systems. In this case we have to replicate the sensing and fusing capabilities of human pilots. Sensors are also expensive. We want to give the autonomous UAVs human-like capabilities without huge costs. Biomimetics concepts such as "optical flow" [Ref. 4] are being put to good use today, and other concepts could help with sensor affordability.

- Social insects communicate directly, via the environment (pheromone trails, etc.), or via the task. These are two examples of stigmergy [Ref. 4]. One of the big challenges of autonomous ops of multi-UAV packages is to minimize communications [Note 4]. Stigmergent techniques promise to reduce bandwidth, but can they be used for flight critical applications?

Artificial Life Artificial Life (ALife) is closely related to biomimetics. Simply put, it's the development of technology that mimics the results of nature (while biomimetics is literally doing it as nature does it). ALife concepts could help us in several areas, reducing the complexity of inner control loops, reducing software size, increasing robustness, and building teams.

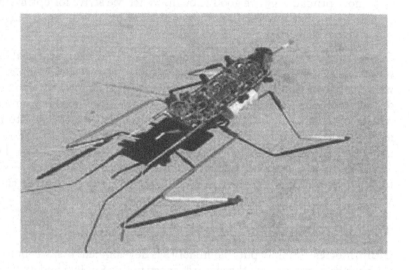

Figure 3.4. Simple, Robust, Analog Robots

- Reducing Control Complexity. The simpler we make control laws, the less processing power required for control, and the more available for processing intensive items, such as automatic target recognition. Some interesting research in this area is being done by Los Alamos National Laboratory. Figure 3.4 shows a crawling robot based simple analog technology that exhibits robust behavior [Ref 5]. It is unclear if this, and other ALife technologies, can be extended to large flying systems. However, if practical, the control algorithm size reduction realized could be enormous.

- Behavior based systems have proven their capability to operate well under uncertainty using simple algorithms [Ref 6]. The tasks may not be done optimally, but they are "good enough", and that is what we desire. Basing our systems on behaviors, rather than functions, gives us greater robustness while minimizing code. The "trick" is to capture the range of behaviors required to accomplish a task.

- Emergent Behavior is when a group of entities with their own simple behaviors (micro-behaviors) interact to make a group behavior. This group behavior, such as fish schooling, is not explicitly coded, but emerges implicitly from individual interactions. If we could do this for a group of UAVs we can eliminate large blocks of code, but several roadblocks exist. How do we predict a priori the wanted behaviors without getting unwanted behaviors? How to we flight certify behaviors which can be non-deterministic?

- Self-Organization is where a group of individuals accomplish a goal without being told how to do it, i.e., no detailed master plan. Social insect colonies are excellent examples. Nowhere in a termite's DNA is the explicit code to build a nest. The implicit interactions of emergent behaviors result in the nest. This can reduce both code (no explicit models) and communication (no global information needed) requirements, but it presents even a trickier design and certification problem than simply emergent behavior.

Evolutionary Computation   Allowing computers to use evolutionary rules to develop algorithms has been used for many years with good results. We want to leverage the technology in a couple of areas.

- Testing software is a costly process. As stated before, software development costs currently approach $ 500 per line of code due to code complexity. Testing million-line software loads without reductions in manpower requirements won't do anything for our affordability concerns. In our attempt to hold the line on development and testing costs we need to develop new methods, not just expand on old ones. We need to not just automate testing, but to slash the man-hours burned in changing software. To do this we must develop tools that will know how to ferret out software problems. Researchers have already started to do this, using evolutionary computational methods to successfully verify and validate autonomous control code [Ref 10]. We need to extend this to flight critical software.

- Autonomous System Development, especially emergent behaviors, will only be practically be enabled through evolutionary methods. Currently we design these via time intensive trial and error processes. It makes much more sense to let a computer evolve the proper sets of behaviors to accomplish tasks. We need to develop the algorithms required for computers to do this [Note 5]. We need to breed, not develop the best systems. Researchers are already accomplishing self-designing, self-constructing robots this way, we need to do this for autonomous UAV control algorithms [Ref 1].

Science Fiction   Believe it or not, we are turning to science fiction writings to help our technology development, especially in the area of safety. Broadly stated, our goal is to develop systems that not only do what we want them to, but also will not hurt the operators and minimize the risk to innocent bystanders. We do not want to hurt friendlies. We have modified Isaac Asimov's Robot Laws [Ref 8] for our case as shown in Figure 3.5.

<u>**First Law:**</u> A flyborg may not injure a 'friendly', or, through inaction, allow a 'friendly' to come to harm.

<u>**Second Law:**</u> A flyborg must obey directions given it by 'friendlies', except where such orders would conflict with the First Law.

<u>**Third Law:**</u> A flyborg must protect its own existence as long as such protection does not conflict with the First or Second Law.

Figure 3.5.   **Flyborg Laws**

The First Law is the safety requirement. We want actuator monitors assuring equipment failures won't result in the UAV crashing into Aunt Bertha's house, automatic collision avoidance algorithms insuring the UAV won't hit manned aircraft, and the self-situational awareness never mistaking a refugee bus as a T-72.

The Second Law can be looked at as eliminating "friendly fire". It is also the sanity check to avoid My Lai massacres and other tragedies.

The Third Law can be considered threat or obstacle avoidance (as long as the obstacle doesn't have friendlies on it, then see First Rule). As with the robotics law, the Third Law might result in UAVs sacrificing themselves to save friendlies.

The "Flyborg Rules" should be looked at as the "Meta Rules" that bound flyborg behaviors. More about Meta rules later in this paper

Other Finally, we believe that to design autonomous control algorithms for distributed coordinated UAV control as a holistic group control problem. We must design for all vehicles simultaneously, not one vehicle at a time (very similar to the discussion of Figure 3.2). Those researching multi-agent telerobotics [Ref 7] share this view. What is needed is a control development package that let's the designers develop control laws for all the members of the multiple-UAV group simultaneously, and will allow them to discriminate particular code changes to group behavior, using evolutionary computation to breed the best behaviors.

Putting it all together:

Eventually we want to enable the technology to build a control systems as in Figure 3.6, using the simple ALife technologies emerging today to run the inner loops, whereas the outer loops would be emergent, self-organizing systems.

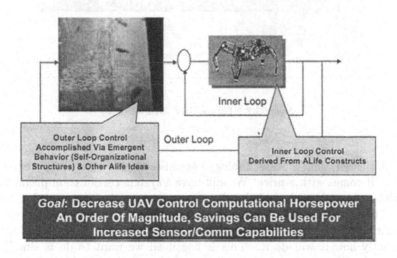

Figure 3.6. Bio-Inspired UAVs

The overall goal of doing this is to decrease the computational requirements at least an order of magnitude below current AI techniques, theoretically reducing costs similarly. In doing this we may have simplified our software; however, we have developed a potential non-deterministic system. In order to ensure performance and safety we need to "wrap" deterministic "meta-rules" around the control loops. This is shown in Figure 3.7, with the Flyborg Rules representing one possible set of meta-rules. We allow the inner loops to be the control system of Figure 3.6, simple, possibly analog, designed along the lines of natural systems. We then use deterministic meta-rules to bind the system behaviors to our bidding.

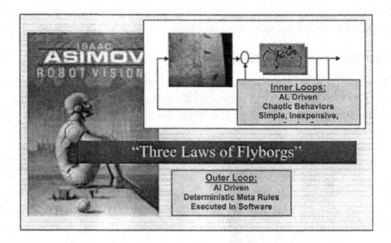

Figure 3.7. Future UAV Architectures

Although using these technologies promises to relieve our software burden, it comes with a price. We still have a system certification problem. Although we have bound the non-determinism with concrete rules, we still have parts of our system that is non-deterministic. Essentially we've built a system where we know what the system will do, but we can't say exactly how it will do it. This is fine if all we want to do is simulations, but if we want to place this aboard large flying vehicles in manned airspace we run into regulations that require determinism at all levels. What we require are methods to prove that although parts of the system may be non-deterministic, the system (or system of systems for multiple UAVs) is deterministic at the levels that count. We have started to look into the techniques for doing this, and it will be a top research goal for our area in the future.

5.   Where are we heading from here?

Our technical focus is currently heavily vested in ordinary AI system building - embedding knowledge through "if-then" rules in finite-state machines. As stated previously, this approach is yielding large blocks of complicated code. To counter this we've started working behavior-based neural nets, and have had good success, transitioning some of the technology. Along with this we've also started looking at formal methods for validating and verifying non-deterministic software.

This is not enough, and in recognition of that fact we are moving into biomimetics, ALife, evolutionary computation, and other non-traditional intelligence building technologies, striving for teaming across the Department of Defense and our world partners. Plans have been made and are being implemented. In order to realize autonomous multi-UAV control systems in practical applications, greater emphasis has to be placed on safety and affordability. We believe that the emerging nature-based technologies coupled with a little science fiction give us the tool to realize our goals.

References

[1] The Wizard Of Oz. Metro-Golden-Mayer, 1939.

[2] Clough, B. Autonomous UAV Control System Safety - What Should It Be, How Do We Reach It, And What Should We Call It? NAE-CON 2000 paper. October 2000

[3] An average from numerous discussions with Department of Defense researchers and industry experts.

[4] Cooper, B. Introduction to Optical Flow, http://ciips.ee.uwa.edu.au/~bruce/research/talk/sld001.htm, University of Western Australia.

[5] Hasslacher, B., Tilden, M. "Living Machines", Robotics and Autonomous Systems: The Biology and Technology of Intelligent Autonomous Agents. Elsevier Publishers, Spring 1995 (Los Alamos Report LAUR-94-2636)

[6] Arkin, R., et al. "Behavioral models of the praying mantis as a basis for robotic behavior". Robotics and Autonomous Systems 32. Elsevier Publishers, 2000 (Cited for the excellent references it contains on ALife studies)

[7] Mataric, M. Coordination and learning in mutirobot systems. IEEE Intelligent Systems Magazine, Vol. 13, No. 4, March/April 1998. pp. 6-8

[8] Asimov, I. Runaround. Astounding Science Fiction. 1942

[9] Asimov, I. Robot Visions. Penguin Books. 1990

[10] Schultz, A. Adaptive Testing: Learning To Break Things. Available in the Web from www.aic.nrl.navy.mil/~schultz/research/adap-testing/index.html . 1999

[11] Lipson, H., and Pollack, J. "Automatic design and manufacture of robotic lifeforms". Nature, Macmillan Magazines Ltd., Vol 406, 31 August 2000.

Appendix: Notes

1 Some people might think that safety is the number one reason. That might be true currently, but a long term USAF goal, to put it simple, is to do more missions cheaper, and cheaper means to use less people less often. Autonomous systems enable significant manpower reductions which lead to cost savings.

2 Some folks have the faulty notion that since we'll go to dual redundant systems with UAVs that we'll save a lot of cost over the triplex and quad systems found in manned aircraft. If one looks at current dual systems, the increased software required to do fault detection and isolation counters the hardware cost reduction. To make dual systems as reliable as triplex systems we are going with a virtual third channel which adds even more software costs. You do not get something for nothing.

3 The Air Force Research Laboratory has several efforts underway in autonomous control of a multi-UAV strike packages, and what we're finding out is that if one includes both inner and outer loop software along with the health management, operating system, and middle-ware expected on future open systems architecture, the whole thing could be pushing a million lines of code.

4 We are also looking at net-based methods such as publisher-subscriber networks to reduce required communication bandwidths. Right now most multi-UAV control concepts rely on global data sent via point to point communication, the highest-bandwidth, most complicated communication scheme possible.

5 There are two schools of thought on developing true machine intelligence. One is that humans can learn to code this by hand, the other is that only computers will be able to teach computers how to think, and that process will heavily involve evolutionary computation. We tend to think that the second school is probably right, but then again, we're only applications engineers.

6 6. By exchanging flight critical data, such as position and velocity for deconfliction and collision avoidance, we are now exposing our critical data links to electronic combat and information warfare. Our designs need to take this into account so antagonists cannot jam or hack our systems.

Chapter 4

USING GRASP FOR CHOOSING BEST PERIODIC OBSERVATION STRATEGY IN STOCHASTIC SYSTEMS FILTERING *

Paola Festa and Giancarlo Raiconi
Department of Mathematics and Informatics
University of Salerno
via S. Allende
84081 Baronissi (SA) Italy.
{gianni,paofes}@unisa.it

Abstract   The problem of optimal periodic scheduling of single channel measures for the state estimation of a multi output discrete time stochastic system is considered. The optimality criterion chosen is the value of the trace of the error covariance matrix of Kalman filter in the periodic steady state, averaged over the observation period. Two interesting examples for practical applications, are studied. The first one considers the case of a number of independent single output subsystems observed by a single observation channel, while the second case deals with the optimization of measurement points and of the relative scanning sequence for the model of a parabolic distributed parameter system. Given the combinatorial nature of the resulting problem, an approximate global optimization method is used to solve it and heuristic rules are devised to overcome difficulties arising from possibly slow convergence in computation of objective function. Numerical examples are reported showing a great improvement with respect to the standard scanning policy.

Keywords: GRASP, filtering, stochastic system

*Work funded by Italian Ministry of University and Scientific Research (project COSO)

R. Murphey and P.M. Pardalos (eds.), Cooperative Control and Optimization, 55–72.
© *2002 Kluwer Academic Publishers.*

1.　　Introduction and problem statement

The optimization of the output transformation in order to minimize a measure of the error covariance in state estimation problems has received some attention in literature ([3],[6],[12],[15]), mainly with reference to the location of sensors in distributed parameter systems (DPS) ([1],[5],[13], [14],[16]). Some authors assume that the output transformation is time invariant and minimize either the steady state estimation error variance, or an averaged value of the expected RMS error over a fixed time duration or else the same function evaluated at a specified terminal time. In some papers, the output operator is assumed time varying, for instance to model a moving sensor. In [7] the problem is formulated as time varying output operator over an infinite time horizon and a suboptimal sequential selection procedure is implemented that, in most cases, leads to periodic measurement policy. In [8] and [9] the periodic constraint is imposed *apriori* searching for the best periodic observation scheme with respect to an averaged measure of the estimation covariance in the periodic steady state. This kind of approach is used to select the observation points among all admissible ones and to find the best time schedule for the selected measurement points.

The optimization procedure suggested in those papers is based on local searches starting from a randomly chosen sequence. Unfortunately, given the non-convexity of the problem, there is no guarantee that this type of algorithms finds a globally optimal solution. In this paper, the optimal solution is searched using a GRASP approach, which is one of most effective global optimization techniques proposed in latest years for solving difficult combinatorial problems.

Generally speaking, the problem is formulated as follows: Let us consider a single output discrete time linear stochastic system characterized by time invariant realization matrices except, for the output vector $c(k) \in \Re^n$ which is time varying:

$$
\begin{aligned}
x(k+1) &= Ax(k) + w(k), & (1) \\
y(k) &= c^T(k)x(k) + v(k), & (2)
\end{aligned}
$$

where $A \in \Re^{n \times n}$ is the dynamic matrix, $w(k)$ is a vector valued zero mean stationary white noise with covariance Q, and $v(k)$ is a real valued zero mean stationary white noise, with intensity r uncorrelated with $w(k)$. The initial state $x(k_0)$ is zero mean random vector of finite covariance P_0 uncorrelated with $w(k)$ and $v(k)$.

The Kalman filter algorithm for the estimation of the system state $x(k+1)$ is expressed as:

$$
\begin{aligned}
\hat{x}(k+1) &= [I - K(k+1)c^T(k+1)]A\hat{x}(k) + K(k+1)y(k+1), \\
&\quad \hat{x}(k_0) = 0, \\
K(k) &= P(k)c(k)/r, \\
P(k+1) &= S(P(k), c(k+1), A, Q, r), \quad P(k_0) = P_0,
\end{aligned}
$$

where the matrix function S is defined as:

$$
S(P, c, A, Q, r) = APA^T + Q - \frac{(APA^T + Q)cc^T(APA^T + Q)}{c^T(APA^T + Q)c + r}. \tag{3}
$$

If the system is detectable in time varying sense [2], the covariance $P(k)$ of the filtered estimate $\hat{x}(k)$ is bounded from above and the dynamic matrix

$$
[I - K(k+1)c^T(k+1)]A
$$

of the filter is exponentially stable. Using the results obtained in [4] it can be shown that if $c(.)$ is a $N-$periodic sequence and $(A, c(.))$ is detectable, then as $k_0 \to -\infty$ the estimation error converges to a ciclostationary zero mean random sequence and its covariance matrix is a symmetric N-periodic positive semi definite (SPPSD) solution of the difference Riccati equation with boundary condition $P(0) = P(N)$.

In the paper, the optimality criterion J used is the mean value in the period of the trace of this SPPSD solution, because if the detectabiliy assumption fails the Riccati equation does not converge, and the feasible set must be restricted to $N-$periodic sequences $c(.)$ such that pairs $(A, c(.))$ are detectables. Conversely, the definition of objective function can extended by putting $J = +\infty$ in the non-detectable case.

Moreover, let us assume that at any time the output vector c can assume one of q prescribed values, that is:

$$
c(k) \in H = \{h_1, ..., h_q\}, \forall k \; h_i \in \Re^n, i = 1, 2, ..q.
$$

Then, the mathematical statement of the problem is the following:

Problem 1 Find a $N-$ple $(c_1^*, c_2^*, ..., c_N^*) \in H^N$ such that:

$$
J(c_1^*, c_2^*, ..., c_N^*) \leq J(c_1, c_2, ..., c_N), \quad \forall (c_1, c_2, ..., c_N) \in H^N,
$$

where :

$$J\left(c_1, c_2, ..., c_N\right) = \left\{ \begin{array}{ll} \frac{1}{N} \sum_{i=0}^{N-1} \mathrm{Tr}(P(i)) & \text{if } (A, c(.)) \text{ is detectable,} \\ +\infty & \text{otherwise,} \end{array} \right.$$

and $P(i)$ satisfies the difference Riccati equation, with two point boundary conditions:

$$P(i+1) = S(P(i), c_{i+1}, A, Q, r), \ i = 0, 1, ..., N-1,$$
$$P(0) = P(N) \geq 0. \tag{4}$$

Since the feasible set is finite the problem could be solved by comparing all the q^N sequences in the set H^N. It is useless to say that this task can be extremely bundersome, because of the high dimension of the feasible set itself in practical instances and because of the computational cost needed for evaluating the objective function. In fact, each evaluation involves the computation of a SPPSD solution of Riccati equation, that can be obtained by iterating the Riccati equation starting from an arbitrary semidefinite positive initial condition until convergence to the SPPSD solution is attained. The convergence is assured under detectability hypotheses but it can be extremely slow. On the other hand, the detectabiliy test itself is very cumbersome. Fortunately, detectable instances that give rise to very slow convergence tend to give even very bad measures for the optimality criterion. Therefore, a test based on the first few iterations of the Riccati algorithm can indicate a rapidly growing objective function can identify detectable sequences not candidate to be optimal or even non detectable sequences.

2.   Two interesting particular cases

In this section, two cases particularly interesting for the applications are described. The first case regards the problem of acquiring and processing a number of independent noise corrupted data streams by means of a single time shared information channel, while the second one is the problem of combined optimal location of sensors and scanning policy for a linear DPS forced by disturbances modeled by random fields.

2.1.   The case of q independent subsystems

Any of the data streams are supposed well modelled by a single output linear discrete time stochastic system. Input noise acting on a system is assumed to be uncorrelated the others. In this case the target is to filter the state of a set of q different stochastic systems by sampling their outputs, acting a multiplexer that can acquire only the output value of a system at any time. In Figure 4.1 the switching scheme is of depicted.

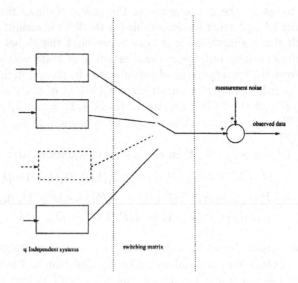

Figure 4.1. observation switching scheme

The trivial control policy of the multiplexer shares uniformly the observation time among systems by sequentially scanning all outputs.

Given different characteristics of systems (stability, observability and noise intensities), it is more interesting to find an *ad-hoc* scanning policy that privileges the observations of some systems with respect to the others. This can be embedded into the framework described in previous section, by simply specializing the definition of the matrices involved in the model (1.1,1.2) as:

$$A = \begin{bmatrix} A_1 & 0 & \dots & 0 \\ 0 & A_2 & \dots & 0 \\ 0 & 0 & \dots & 0 \\ 0 & 0 & \dots & A_q \end{bmatrix}, Q = \begin{bmatrix} Q_1 & 0 & \dots & 0 \\ 0 & Q_2 & \dots & 0 \\ 0 & 0 & \dots & 0 \\ 0 & 0 & \dots & Q_q \end{bmatrix},$$

$$P_0 = \begin{bmatrix} P_{01} & 0 & \dots & 0 \\ 0 & P_{02} & \dots & 0 \\ 0 & 0 & \dots & 0 \\ 0 & 0 & \dots & P_{0q} \end{bmatrix}, h_i = \begin{bmatrix} 0 \\ \dots \\ g_i \\ \dots \\ 0 \end{bmatrix} \quad i - \text{th block},$$

$A_i \in \Re^{n_i \times n_i}$, $Q_i = Q_i^T \in \Re^{n_i \times n_i}$, $P_{0i} = P_{0i}^T \in \Re^{n_i \times n_i}$, $g_i \in \Re^{n_i}$, $i = 1, 2, \dots, q$.

In order to assure the convergence of the global Kalman filter algorithm all pairs (A_i, g_i) must be detectable (in the time invariant systems sense). With this assumptions it is easy to see that the global filtering algorithm gives nothing but the optimal estimate of state vector of any of the q subsystems respect to the observed data. In more detail, at time $k + 1$ the algorithm computes an estimate $\hat{x}_i(k + 1)$ of state subvector $x_i(k + 1) \in \Re^{n_i}$ of the i–th subsystem. If $c(k + 1) = h_i$, $\hat{x}_i(k + 1)$ is the optimal filtered estimate, otherwise it is the optimal linear predictor based on $\hat{x}_i(k)$.

Formally, let $c(k + 1) = h_i$ then the updating equations are:

$$\hat{x}_i(k + 1) = [I - K_i(k + 1)g_i^T(k + 1)]A_i\hat{x}_i(k) + K_i(k + 1)y(k + 1),$$
$$K_i(k) = P_i(k)g_i(k)/r, \ P_i(k + 1) = S(P_i(k), g_i(k + 1), A_i, Q_i, r),$$
$$\hat{x}_j(k + 1) = A_j\hat{x}_j(k), \ P_j(k + 1) = A_jP_j(k)A_j^T + Q_j, \ j \neq i.$$

Lower and upper bounds computation..   For designing approximate or exact combinatorial optimization algorithms, it is useful to *apriori* have lower or upper bounds on the objective function values to discard candidate solutions when implementing local search procedures or for judging about the tightness of approximate solutions found.

As the lower bound is concerned, in the present case the following proposition can be helpful:

Proposition 2.1 *Let $\overline{P}_i, i = 1, 2, ..., q$ be the symmetric semidefinite positive solution of the discrete time algebraic Riccati equation $\overline{P}_i = S(\overline{P}_i, g_i, A_i, Q_i, r)$ (which must exist given the detectability assumption). Then, $\forall N$ and $\forall (c_1, c_2, ..., c_N) \in H^N$ it holds that*

$$J(c_1, c_2, ..., c_N) \geq LB = \sum_{i=1}^{q} Tr(\overline{P}_i).$$

Proof $\overline{P}_i$ is the covariance of the steady state Kalman filter for the subsystem characterized by matrices (A_i, g_i, Q_i, r). Given any periodic sequence $(c_1, c_2, ..., c_N)$, the steady state covariance of the estimate for the state of i–th subsystem (i-th diagonal block of the SPPSD matrix $P(k)$) is a periodic matrix $P_i(k)$ and is obtained by applying a modification of the Kalman filter rule for the i–th subsystem (the Kalman gain is zero when the subsystem is not observed). By optimality of Kalman filter, $Tr(\overline{P}_i) \leq Tr(P_i(k))$ that completes the proof.   ∎

In order to find an upper bound on the optimal objective function value, any admissible input can be chosen. If N is either free or fixed but multiple of q, an upper bound can still computed in terms of the solution of some algebraic Riccati equations, as stated by the following proposition.

Proposition 2.2 *Let* $\overline{\overline{P}}_i$, $i = 1, 2, ...q$ *be the symmetric positive semidefinite solution of the discrete time algebraic Riccati equation* $\overline{\overline{P}}_i = S(\overline{\overline{P}}_i, g_i,$ $A_i^q, \overline{Q}_i, r)$, *where* $\overline{Q}_i = \sum_{j=0}^{q-1} A_i^j Q_i \left(A_i^j\right)^T$. *Then, for any* N *multiple of* q, *an upper bound for the optimal value of the objective function is:*

$$UB = \frac{1}{N} \sum_{i=1}^{N} \sum_{k=0}^{q} \left(A_i^k \overline{\overline{P}}_i \left(A_i^k\right)^T + \sum_{j=0}^{k-1} A_i^j Q_i \left(A_i^j\right)^T \right).$$

Proof The thesis is easily proved by explicitly computing the objective function for the scanning sequence $c(k - lN) = h(k)$, $k = 1, 2, ..., N$,l integer. Note that by definition, UB is the value of the objective function corresponding to the cyclic scanning of system outputs, which does not depend from the *order* of the scanning itself. ∎

2.2.   A Distributed Parameter System problem.

Following [9] , let us consider a stochastic parabolic DPS described by

$$\frac{\partial u}{\partial t} = \Theta u + w, \tag{5}$$

where the state of the system is represented by $u(z, t)$, $t > t_0$, $z = (z_1, z_2) \in \Omega \subset \Re^2$, while Θ is a second order partial differential operator of elliptic type over a bounded spatial domain Ω with appropriate boundary conditions assigned on $\Gamma = \partial\Omega$

$$L(u, \frac{\partial u}{\partial n}) = f(z, t), \; z \in \Gamma = \partial\Omega, t \in [t_0, +\infty).$$

The boundary data f depends linearly from an exponentially correlated stochastic process $b(t)$.

The initial conditions are

$$u(z, t_0) = u_0(z), z \in \Omega.$$

The input noise $w(z, t)$ is a zero mean, spatially distributed, white in time random function characterized by the covariance function $E(w(z, t),$ $w(\zeta, \tau)) = q(z, \zeta)\delta(t - \tau)$. The state of the system is measured at discrete time istants, uniformly spaced by sampling interval T. At each time, the value of the state $u(z, t)$ can be known. Output data are corrupted by a white measurement error and the output equation is

$$y(k) = u(kT, z(kT)) + \nu(k). \tag{6}$$

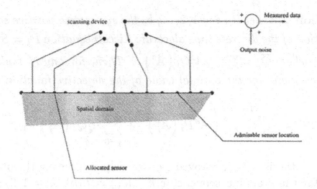

Figure 4.2. Scanning observation of a Distributed Parameter System.

In this case, the target is to find the optimal measurement policy that consists in choosing the best observation points and their optimal periodic scanning.

Under suitable hipoteses about operators Θ and L, and if boundary and initial data are sufficiently regular, a solution of (2.1) can expressed as $u(t,z) = u^0(t,z) + u^P(t,z)$, where u^P satisfies the homogeneous equation $\Theta u^P = 0$ with boundary conditions $L(u^P, \frac{\partial u^P}{\partial n}) = f(t,z) \, \forall t > t_0$ and u^0 the inhomogeneous equation:

$$u_t^0 = \Theta u^0 - u_t^P + w, \, t > 0, \, z \in \Omega. \tag{7}$$

Initially,

$$u^0(t_0, z) = u_0(z) + u^P(t_0 z) \, z \in \Omega, \tag{8}$$

and homogeneous boundary conditions:

$$L(u^P, \frac{\partial u^P}{\partial n}) = 0, \, t > 0, \, z \in \Gamma. \tag{9}$$

Using a modal approach, equations (2.3-2.5) can be formulated in terms of orthonormal eigenfunction of the elliptic operator $\{\varphi_i(z), z \in \Omega, i = 1, 2, ...\}$. The forcing terms of (2.2) u^P and w can expanded in Fourier series as follows:

$$u^P(t,z) = \sum_{i=0}^{\infty} u_i^P(t)\varphi_i(z),$$

$$u_i^P(t) = \int_\Omega u^P(t,z)\varphi_i(z)d\Omega = d_i b(t),$$

$$w(t,z) = \sum_{i=0}^{\infty} w_i(t)\varphi_i(z),$$

$$w_i(t) = \int_\Omega w(t,z)\varphi_i(z)d\Omega.$$

Let λ_i, $i = 1, 2...$ be the eigenvalue associated to φ_i, then for any t, the solution u^0 of (2.3-2.5) can expressed as the convergent (in $L_2(\Omega)$ sense) series:

$$u^0(t, z) = \sum_{i=1}^{\infty} u_i^0(t)\varphi_i(z),$$

where Fourier coefficients u_i^0, $i = 1, 2, ...$ satisfy the stochastic differential equation:

$$\frac{du_i^0(t)}{dt} = \lambda_i u_i^0(t) - d_i \frac{db(t)}{dt} + w_i(t). \tag{10}$$

Solving the ordinary differential equations (2-6) over an interval between two measurement instants and recalling that $u_i(t) = u_i^0(t) - u_i^p(t)$, we obtain that

$$u_i((k+1)T) = e^{\lambda_i T} u_i^0(kT) - \int_0^T e^{\lambda_i(T-s)} d_i \frac{db}{dt}(kT + s)ds +$$
$$\int_0^T e^{\lambda_i(T-s)} w_i(kT + s)ds + d_i b((k+1)T),$$

Applying integration by parts of second term gives:

$$u_i((k+1)T) = e^{\lambda_i T} u_i(kT) - d_i\lambda_i \int_0^T e^{\lambda_i(T-s)} b(kT + s)ds +$$
$$\int_0^T e^{\lambda_i(T-s)} w_i(kT + s)ds.$$

Finally, by dropping the sampling time T into the arguments, one obtains the following discrete time system rappresentation.

$$u_i(k+1) = \mu_i(k)u_i(k) + \omega_i(k), \tag{11}$$

with $\mu_i = e^{\lambda_i T}$, and

$$\omega_i(k) = -d_i\lambda_i \int_0^T e^{\lambda_i(T-s)} b(kT + s)ds + \int_0^T e^{\lambda_i(T-s)} w_i(kT + s)ds.$$

The output equation (2.2) can be rewritten as

$$y(k) = \sum_{i=0}^{\infty} \varphi_i(z(kT))u_i(k) + v(k).$$

Finite dimensional model. The discrete time system and output equations (2.6,2.7) are similar to equations (1.1,1.2), but infinite dimensional. A finite dimensional approximation of the system can obtained by considering only the first n modes of the distributed parameter operator, that is by truncating the series to n-th term and the approximate state $u^n(kT, z)$ is given by

$$u^n(kT, z) = \sum_{i=1}^{n} x_i^n(k)\varphi_i(z),$$

where x_i^n is the i-th component of vector u^n that satisfies the following difference equation:

$$x^n(k+1) = M^n x^n(k) + \omega^n(k), \tag{12}$$

where:

$$M^n = \text{diag}(\mu_1, \mu_2, ..., \mu_n), \omega^n(k) = [\omega_1(k), \omega_2(k), ..., \omega_n(k)]^T.$$

Analogously, the approximate output equation becomes

$$y(k) = (c^n(k))^T x^n(k) + v(k), \tag{13}$$

with $c^n(k) = [\varphi_1(z(kT)), \varphi_2(z(kT)), ..., \varphi_n(z(kT))]^T$. Once q points $z_1, z_2, ..., z_q$, on the spatial domain Ω are preselected as candidate locations for positioning sensor devices, the set of admissible output vectors is defined as well:

$$H = \left\{ \begin{bmatrix} \varphi_1(z_1) \\ \varphi_2(z_1) \\ ... \\ \varphi_n(z_1) \end{bmatrix}, \begin{bmatrix} \varphi_1(z_2) \\ \varphi_2(z_2) \\ ... \\ \varphi_n(z_2) \end{bmatrix}, ..., \begin{bmatrix} \varphi_1(z_q) \\ \varphi_2(z_q) \\ ... \\ \varphi_n(z_q) \end{bmatrix} \right\}, \tag{14}$$

and state equation (2.8), output equation (2.9), and the admissible set (2.10) completely define data for *Problem 1.1*.

Note that in the last considered case there are several structural differences with respect to that considered above. Even if system matrix is diagonal, all modes are linked together both because matrix Q is in general full and also because all entries of output vectors in (2.10) are generally different from zero. Morever, in the case of previous section the scope is to find the best way to observe q different processes, possibly some more frequently than others, but *all processes must be observed* to say something about their status. Therefore, it is natural to assume $N > q$ and even N a multiple of q. Instead, in the present case $z_1, z_2, ..., z_q$

describe all the admissible places where sensor can be located. Since the actual number of used sensors is in general less than q, it is reasonable assume $N < q$. The optimization procedure serves in this case primarily to decide sensor locations.

About detectability issues, in this case one can observe that it is guaranteed because of the exponential stability of the system ($\mu_i < 1, \forall i$). Moreover, the complete observability of a pair (M^n, c^n) is assured if all entries of $c^n = [\varphi_1(z), \varphi_2(z), ..., \varphi_n(z)]^T$ are not zero (i.e. if z does not fall in a zero of one of first n eigenfunctions). This can assured by a proper choice of the observation set.

Even in this case an upper bound on the optimal objective function can be obtained by considering the steady state filter for the optimally located single fixed observation point.

3.   Discrete problem formulation

The discrete nature of the problem allows a formulation characterized by zero-one decision variables $\alpha_{ij} \in \{0,1\}, i = 1, 2, .., q, j = 1, 2, ...N$ subject to constraits $\sum_{i=1}^{q} \alpha_{ij} = 1$. Expressing the output vector as $c(k) = \sum_{i=1}^{q} \alpha_{ij} h_i, k = j + \nu N, \nu = 0, 1, ...$, the new problem formulation is:

Problem 2 Let $\alpha = \{\alpha_{ij}\} \in \{0,1\}^{q \times N}$. Then find $\widehat{\alpha} = \{\widehat{\alpha}_{ij}\}, i = 1, 2, .., q, j = 1, 2, .., N$ such that:

$$J(\widehat{\alpha}) \leq J(\alpha) \, \forall \alpha \in \left\{ \{0,1\}^{q \times N} \mid \sum_{i=1}^{q} \alpha_{ij} = 1, j = 1, 2, ..., N \right\},$$

where $J(\alpha)$ is defined as

$$J(\alpha) = \lim_{m \to \infty} J_m = \lim_{m \to \infty} \frac{1}{N} \sum_{j=1}^{N} Tr(P_j + mN),$$

$$P_{mN+j+1} = \left(AP_{mN+j} A^T \right) - \sum_{i=1}^{q} \alpha_{ij} \frac{\left(AP_{mN+j} A^T \right) h_i h_i^T \left(AP_{mN+j} A^T \right)}{r + h_i^T \left(AP_{mN+j} A^T \right) h_i},$$
$$P_0 = 0.$$

The conditions assuring the convergence of the covariance matrix to a unique (definite positive) periodic sequence are supposed verified[1]. Since the initial condition is arbitrary, without loss of generality we can assume that $P_0 = 0$. If for a certain α the system is not detectable, then the definition of J does not converge and $J(\alpha) = +\infty$ is assumed. From a practical point of view, two possible alternatives are open to deal with this case: the implementation of a cumbersome algebraic detectability test or a test applied directly to convergence of P. We use the second approach, even for overcome computational difficulties arising when scarcely detectable solutions are encountered. Convergence of J_m may be extremely slow, especially in instances with poor detectability structure, which generally leads to high value of J. Since the index must be minimized, a lot of computation time is saved if we are able to foresee that a sequence is unlikely to be better than another one. Let us consider the sequence J_m obtained starting from $P_0 = 0$. Usually, the convergence of $J_n \to J$ becomes from below, as soon as $J_m(a)$ becomes grater than a known upper bound (or greater than the value of the current solution during the local search phase). In this case, it is reasonable to not further consider α as possible solution even if convergence is not reached.

3.1.   A GRASP for global optimization

A *Greedy Randomized Adaptive Search Procedure* (GRASP) is a metaheuristic for finding approximate solutions for difficult combinatorial problems. GRASP is a multistart method characterized by two phases: a construction phase and a local search also known as local improvement phase. During the construction phase a feasible solution is iteratively constructed. One element at time is randomly chosen from a *Restricted Candidate List* (RCL), whose elements are sorted according to some greedy criterion, and added to the building admissible solution. As the found solution could not be locally optimal with respect to the adopted neighborhood definition, the local search phase tries to improve it. These two phases are iterated and the best solution found is kept as an approximation of the optimal one.

GRASP has been proposed in 1989 by Feo and Resende in [10]. Since 1989 numerous papers on the basic aspects of GRASP, as well as enhancements to the basic metaheuristic have been appeared in the literature. GRASP has been applied to a wide range of combinatorial optimization problems, ranging from scheduling and routing to drawing and turbine balancing, as reported in a very recent annotated bibliography of the GRASP literature [11], due to Festa and Resende.

3.2.   Greedy construction phase

At each GRASP construction phase a starting point for the local search procedure is built by following the first N steps of the sequential greedy procedure suggested in [7].

The Kalman filter typical iteration for the covariance is composed of two stages:

- the one step prediction covariance computation:

$$P(k+1|k) = AP(k)A^T;$$

- the filtering covariance updating:

$$P(k+1) = P(k+1|k) - M(k+1)$$
$$M(k+1, c(k+1)) = \frac{P(k+1|k)c(k+1)c^T(k+1)P(k+1|k)}{r+c^T(k+1)P(k+1|k)c(k+1)}.$$

The greedy criterion is the maximization of the trace of term M, which is subtracted to the covariance in the filtering update step.

The GRASP construction phase paradigm requires to choose at any step a member of the RCL defined on the basis of the adopted greedy criterion. In our case, the complete procedure to choose the randomized greedy initial α for a fixed N is the following

Algorithm 3.1 *Construction Phase*
    Step 1: choose a large value for $P(0)$, for example $\frac{1}{\epsilon}I$, set $0 < \lambda < 1$ put $k = 0$
    Step 2: compute $P(k+1|k)$
    Step 3: for $i = 1, 2, ..., q$ compute $b_i = Tr(M(k+1, h_i))$ set $b_m = \min\{b_i\}$, $b_M = \max\{b_i\}$
    Step 4: choose at random $j \in \{1, 2, ..., q\}$ such that $b_M - \lambda(b_M - b_m) \leq b_j$, set $\alpha_{lk+1} = \delta_{jl}$
    Step 5: if $k < N - 1$ compute $P(k+1) = P(k+1|k) - M(k+1, h_j)$, put $k = k + 1$ and go to Step 2, else exit with the matrix $\{\alpha_{ls}\}$ full.

In **step 1** the large value assumed for P_0 respects the no prior knowledge about the initial state. At any iteration, the vectors h_i are sorted according to the value $b_i = Tr(M(k+1, h_i))$ (step 3) and the RCL is constituted only by those vectors j with b_j sufficiently large (step 4).

3.3.   Local search procedure

Once a starting point α is found, it must be verified if it is locally optimal and possibly an improving solution must be searched in a prescribed neighborhood of α. For this pourpose, a distance $\Delta(\alpha, \beta)$ must be introduced in order to measure the similarity of two different candidate solutions α, β. The definition of Δ depends on the physical meaning of relevant parameters. We used the Hamming distance defined as follows.

$$\Delta_H(\alpha, \beta) = \frac{1}{2} \sum_{i=1}^{q} \sum_{j=1}^{N} |\alpha_{ij} - \beta_{ij}|.$$

In our application, Δ_H measures the number of different columns between matrices α and β. Actually, as well as other similarly defined metrics, Δ_H compares two "strings" of length N and is unable to recognize equivalent sequences[2], but this is not a drawback when used for a local search procedure.

For designing an efficient local search phase the size of the neighborhood must be not too large, because this phase is performed iteratively starting from a great number of initial values. In this paper, to maintain a reasonable computational effort, we used the distance Δ_H and neighborhoods of diameter equal to 1 or 2.

3.4.　GRASP

The description of the GRASP used for solving our problem for a fixed N is reported in the following.

Algorithm 3.2 *GRASP*

　Step1: Read systems data. Put: $it = 0$ and $\widehat{J}$ equal to a large number.

　Step 2: Construction: By Algorithm 3.2 *build the RCL and extract a random a starting sequence α_0*

　Step3: Local Search: Find the best value of J on a prescribed neighborhood of α_0. Put: $\alpha^* = \underset{\alpha \in \{\Delta(\alpha, \alpha_0) \leq s\}}{\text{argmin}} \{J(\alpha)\}, \ J^* = J(\alpha^*)$

　Step 4: put $it = it + 1$ if $J^* < \widehat{J}$ put $\widehat{J} = J^*, \widehat{\alpha} = \alpha^*$

　Step 5: if $it = maxit$ stop and return the solution $\widehat{\alpha}$ else go to Step 2.

4.　Numerical results

Several numerical investigationshave been performed in order to verify that the proposed GRASP can obtain satisfying results, which have been very encouraging. In fact, the good solutions have been obtained with a reasonable computational effort. Secondly, since in the proposed approach it is possible to scan multiple observations, the estimation error covariance can be strongly reduced.

In this section, we describe in detail a particular application to the state estimation of five independent linear systems:

Systems data are:

$$A_1 = \begin{bmatrix} 1.01 & 0.8 \\ 0 & 0.9 \end{bmatrix}, Q_1 = \begin{bmatrix} 0.1 & 0.1 \\ 0.1 & 0.1 \end{bmatrix}, g_1 = \begin{bmatrix} 1 \\ 0 \end{bmatrix};$$

$$A_2 = \begin{bmatrix} 0.99 & 0.8 \\ 0 & 0.8 \end{bmatrix}, Q_2 = \begin{bmatrix} 1 & 1 \\ 1 & 1 \end{bmatrix}, g_2 = \begin{bmatrix} 1 \\ 0 \end{bmatrix};$$

$$A_3 = \begin{bmatrix} 0.9501 & 0.4860 & 0.4565 \\ 0.2311 & 0.8913 & 0.0185 \\ 0.6068 & 0.7621 & 0.8214 \end{bmatrix}, Q_3 = \begin{bmatrix} 0.7406 & 1.0186 & 0.8131 \\ 1.0186 & 1.8841 & 1.0338 \\ 0.8131 & 1.0338 & 0.9265 \end{bmatrix},$$

$$g_3 = \begin{bmatrix} 0.1987 \\ 0.6038 \\ 0.2722 \end{bmatrix};$$

$$A_4 = \begin{bmatrix} 0.4892 & 0.8711 \\ 0.6770 & 1.0140 \end{bmatrix}, Q_4 = \begin{bmatrix} 6.6134 & 1.1494 \\ 1.1494 & 0.6040 \end{bmatrix}, g_4 = \begin{bmatrix} 0.1988 \\ 0.0153 \end{bmatrix};$$

$$A_5 = [1.01], Q_5 = [0.1], g_5 = [1].$$

We chosen the period duration N among values $\{10, 15, 20\}$ (which are multiples of $q = 5$). In all experiments the values of the output noise intensity r is equal to 1.0.

GRASP was coded in MATLAB language and then compiled with the MATCOM compiler for Borland C++ Builder in order to obtain a more efficient code.

Figures 4.3 and 4.4 show the best observation schedule found by the algorithm. In the figures a vertical bar denotes a time in which the system is observed. For $N \in \{10, 15, 30\}$ the values obtained for $J(\widehat{\alpha})$ are 308.30, 234.99, and 216.42, respectively. In all cases, a great improvemet with respect to the standard policy value $UB = 837.84$ is obtained. The lower bound was also been computed and is $LB = 41.85$, but a so low value can be accomplished only using hardware components (switcher, AD converter and computing device) five time faster. The behaviors of optimal objective function versus N, together with values of UB and LB are depicted in Figure 4.4.

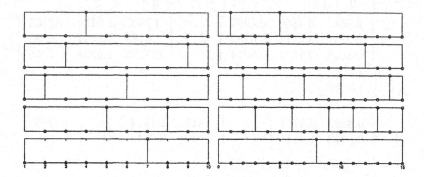

Figure 4.3. Optimal scheduling for $N = 10$ and $N = 15$.

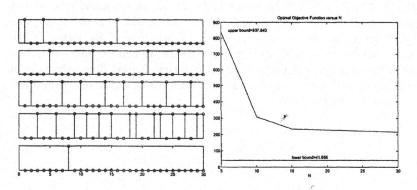

Figure 4.4. Optimal scheduling for $N = 30$ and objetcive function values.

References

[1] Aidarous S. E., Gevers M. R., and Installe M.J.: Optimal sensors allocation strategies for a class of stochastic distributed systems. *International Journal of Control*, 22, (1975), pp. 197-213.

[2] Anderson B.D.O., Moore J.B.: Detectability and stabilizability of time varying discrete time linear systems, *SIAM Journal on Control and Optimization*. 19, (1981), pp. 20-32.

[3] Athans M.: On the determination of optimal costly measurement strategies for linear stochastic systems. *Automatica*, 8 ,(1972), pp.397-412

[4] Bittanti S., Colaneri P., De Nicolao G.: The difference periodic Riccati equation for the periodic prediction problem. *IEEE Transaction on Automatic Control*, 33, (1988), pp. 706-712.

[5] Carotenuto L., Muraca P., Raiconi G.: Optimal location of a moving sensor for the estimation of a distributed parameter system. *International Journal of Control*, 46 (1987), pp. 1671-1688

[6] Carotenuto L., Muraca P., Raiconi G.: On the optimal design of the output transformation for discrete time linear systems. *Journal of Optimization Theory and Applications*, 68, (1991), pp 1-17

[7] Carotenuto L., Muraca P.,Pugliese P., Raiconi G.: Recursive sensor allocation for of distributed parameter system. International Journal of Modelling and Simulation, 12,(1992), pp.117-124

[8] Carotenuto L., Longo G., Raiconi G.: Optimal periodic selection of the measurements for the state estimation of discrete time linear systems. *Proceedings of XII IFAC World Congress* (1993), vol 2, pp. 11-16.

[9] Carotenuto L., Muraca P., Raiconi G.: Periodic Measurement strategies for Distributed Parameters Systems filtering. *Mathematical Modeling of systems*, 1, 4, (1995), pp. 244-260.

[10] Feo T.A. M.G.C. Resende: A probabilistic heuristic for computationally difficult set covering problems. *Operation Research Letters,* 8, (1989), pp. 67-71

[11] Festa P., M.G.C. Resende: GRASP: An Annotated Bibliography, accepted for publication in *Essays and Surveys in Metaheuristics,* Kluwer Academic Publishers, 2001.

[12] Geromel J.C.: Global optimization of measurement strategies for linear stochastic systems. *Automatica,* 21 (1985), pp. 293-300.

[13] Kubrusly C.S., Malebranche H.: Sensors and controllers location in distributed systems: a survey. *Automatica,* 21, (1985), pp. 117-128.

[14] Kumar S., Seinfeld J.H.: Optimal location of measurements for distributed parameter estimation. *IEEE Transaction on Automatic Control,* 23 (1981), pp. 690-698.

[15] Mhera R.K.: Optimization of measurement schedules and sensors design for linear dynamical systems. *IEEE Transaction on automatic Control.* 21, (1976), pp.55-64.

[16] Nakamori Y. Miyamoto S., Ikeda S., Sawaragi Y.: Measurements optimization with sensitivity criteria for distributed parameter systems. *IEEE Transaction of Automatic Control,* 25 (1980), pp.889-901.

Chapter 5

COOPERATIVE CONTROL OF
ROBOT FORMATIONS

Rafael Fierro, Peng Song, Aveek Das, and Vijay Kumar
GRASP Lab. University of Pennsylvania, Philadelphia PA, USA
rfierro, pengs, aveek, kumar@grasp.cis.upenn.edu

Abstract   We describe a framework for controlling and coordinating a group of
nonholonomic mobile robots equipped with range sensors, with applications ranging from scouting and reconnaissance, to search and rescue
and manipulation tasks. We derive control algorithms that allow the
robots to control their position and orientation with respect to neighboring robots or obstacles in the environment. We then outline a coordination protocol that automatically switches between the control laws
to maintain a specified formation. Two simple trajectory generators
are derived from potential field theory. The first allows each robot to
plan its reference trajectory based on the information available to it.
The second scheme requires sharing of information and enables a rigid
group formation. Numerical simulations illustrate the application of
these ideas and demonstrate the scalability of the proposed framework
for a large group of robots.

Keywords: formation control, potential functions, nonholonomic mobile robots,
switching control.

1.   Introduction

It is well known that there are several tasks that can be performed
more efficiently and robustly using multiple robots, see for example [19].
Multi-robot applications include cooperative manipulation, navigation
and planning, collaborative mapping and exploration, and formation control. In fact, there is extensive literature on motion planning and control
of mobile robots in structured environments. However, traditional control theory mostly enables the design of controllers in a single mode of
operation, in which the task and the model of the system are fixed. While

R. Murphey and P.M. Pardalos (eds.), Cooperative Control and Optimization, 73–93.
© *2002 Kluwer Academic Publishers.*

control and estimation theory allows us to model each behavior as a dynamical system, it does not give us the tools to compose behaviors or the hierarchy that might be inherent in the switching behavior, or to predict the global performance of a highly complex multi–robotic system.

The key contributions of this paper are (1) a set of control algorithms and a coordination strategy that allow the robots to maintain a prescribed formation, and (2) a newly developed trajectory generator that combines potential functions and the dynamics of visco-elastic contacts. By combining control, coordination, and trajectory generation we are able to compose single control modes or behaviors and build formations in a modular fashion. Moreover, we can guarantee that under reasonable assumptions the basic formation is stable. Thus, the group can maintain a desired formation and flow towards its goal configuration. The ability to maintain a prescribed formation allows the robots to perform a variety of tasks such as collaborative mapping and exploration, and cooperative manipulation [22].

We divide the multi-robot cooperative control problem into two areas: (a) *formation control* and (b) *trajectory generation*. Formation control approaches can be classified into three main categories as in [4]: *leader-following*, *behavioral* and *virtual structures*. In the leader-following one robot acts as a leader and generates the reference trajectory for the team of robots. Thus, the behavior of the group is defined by the behavior of the leader. In the behavioral approach, a number of basic behaviors is prescribed, *e.g.*, obstacle avoidance, formation keeping, and goal seeking. The overall control action (*emergent behavior*) is a weighted average of the control actions for each basic behavior. In this case, composing control strategies for competing behaviors and implementing them can be straightforward. However, formal stability analysis of the emergent group behavior may be difficult. Finally, virtual structures consider the entire formation as a rigid body. Once the desired dynamics of the virtual structure are defined, then the desired motion for each agent is derived. The framework proposed in this work is flexible enough to accommodate any of these formation control approaches. It is the designer's decision to use decentralized reactive behaviors with no leader involved, leader-following, or rigid body motion to perform a given task. We will demonstrate this through numerical simulation experiments.

The problem of multi–robot trajectory generation is to generate collision free trajectories for mobile robots to reach their desired destinations. Previous approaches in this area can be broadly divided into two classes including graph based planners [3], and potential field methods [14, 15]. In this work we consider the latter. Artificial potential field approaches are based on constructing repulsive potential functions around obstacles and attracting potential functions around the goal location. The design of a potential field with a global minimum at the goal configuration turns out to be difficult. Various techniques have been developed to overcome these difficulties, see for instance [23, 5, 2]. In contrast, we propose the use of simple goal-directed fields that are not specifically designed to avoid obstacles or neighboring robotsas in [20]. Instead, when a robot is close to an obstacle, it adopts a behavior that simulates the dynamics of a visco-elastic collision guaranteeing that the actual collision never occurs. This approach can be potentially scaled to multiple (tens and hundreds) robots and to higher speeds of operation.

The rest of the paper is organized as follows. In section 2 we describe a framework that allows to hierarchically compose *planning* and *control* in a distributed fashion. Section 3 presents the suite of control algorithms and the coordination strategy for switching between these controllers. Then, we formulate the *trajectory generator* in section 4. Section 5 gives simulation results and illustrates the benefits and the limitations of this methodology underlying the implementation of cooperative control of robot formations. Finally, some concluding remarks and future work ideas are given in section 6.

2.   Framework for cooperative control

We describe a framework for decentralized cooperative control of multi-robotic systems that emphasizes simplicity in planning, coordination, and control. The framework incorporates a two-level control hierarchy for each robot consisting of a trajectory generation level and a coordination level as illustrated in Figure 5.1. The trajectory generator derives the reference trajectory for the robot while the coordination level selects the appropriate controller (*behavior*) for the robot.

The availability and sharing of information between the robots greatly influences the design of each level. This is particularly true at the trajectory generation level. The trajectory generator can be completely decentralized so that each robot generates its own reference trajectory based on the information available to it, through its sensors and through the communication network. Alternatively, a designated leader plans its trajectory and the other group members are able to organize themselves to following the leader. The trajectory generators are derived from potential field theory. At the coordination level we assume range sensors that allow the estimation of position of neighboring robots and obstacles. This model is motivated by our experimental platform consisting of mobile robots equipped with omni–directional cameras described in [7, 1]. Each robot chooses from a finite set of control laws that describe

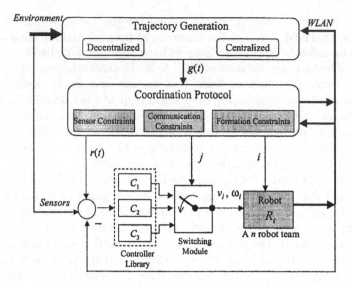

Figure 5.1. A formation control framework.

its interactions with respect its neighbors (robots and obstacles) and allow it to go to a desired goal position. Thus the overall goal of this level is to prescribe the rules of mode switching and thus the dynamics of the switched system [17].

3.   Formation control

In this section, we consider a group of n nonholonomic mobile robots and describe the controllers that specify the interactions between each robot and its neighbor. The robots are velocity controlled platforms and have two independent inputs v_i and ω_i. The control laws are based on I/O feedback linearization. This means we are able to regulate two outputs. Moreover, we assume that the robots are assigned labels from 1 through n which restrict the choice of control laws. Robot 1 is the leader of the group.

We adopt a simple kinematic model for the nonholonomic robots. The kinematics of the ith robot are given by

$$\dot{x}_i = v_i \cos\theta_i, \quad \dot{y}_i = v_i \sin\theta_i, \quad \dot{\theta} = \omega_i \tag{1}$$

where $\boldsymbol{x}_i \equiv (x_i, y_i, \theta_i) \in SE(2)$.

In Figure 5.2, we show subgroups of two and three robots. Robot j is designated as a follower of Robot i. We first describe two controllers, adopted from [9], and derive a third controller that takes into account possible interactions with an obstacle.

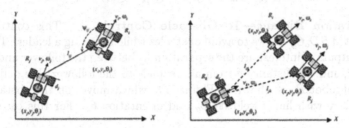

Figure 5.2. The *Separation Bearing* and *Separation Separation* Controllers.

Separation–Bearing Control.   By using this controller (denoted $SB_{ij}C$ here), robot R_j follows R_i with a desired separation l_{ij}^d and desired relative bearing ψ_{ij}^d, see Figure 5.2(left). The control velocities for the *follower* are given by

$$v_j = s_{ij}\cos\gamma_{ij} - l_{ij}\sin\gamma_{ij}(b_{ij} + \omega_i) + v_i\cos(\theta_i - \theta_j) \qquad (2)$$

$$\omega_j = \frac{1}{d}[s_{ij}\sin\gamma_{ij} + l_{ij}\cos\gamma_{ij}(b_{ij} + \omega_i) + v_i\sin(\theta_i - \theta_j)] \qquad (3)$$

where d is the distance from the wheel axis to a reference point on the robot, and

$$\gamma_{ij} = \theta_i + \psi_{ij} - \theta_j, \qquad (4)$$

$$s_{ij} = k_1(l_{ij}^d - l_{ij}), \qquad (5)$$

$$b_{ij} = k_2(\psi_{ij}^d - \psi_{ij}), \quad k_1, k_2 > 0 \qquad (6)$$

The closed-loop linearized system is

$$\dot{l}_{ij} = k_1(l_{ij}^d - l_{ij}), \quad \dot{\psi}_{ij} = k_2(\psi_{ij}^d - \psi_{ij}), \quad \dot{\theta}_j = \omega_j \qquad (7)$$

Separation–Separation Control.   By using this controller (denoted $S_{ik}S_{jk}C$), robot R_k follows R_i and R_j with a desired separations l_{ik}^d and l_{jk}^d, respectively, see Figure 5.2(right). In this case the control velocities for the follower robot become

$$v_k = \frac{s_{ik}\sin\gamma_{jk} - s_{jk}\sin\gamma_{ik} + v_i\cos\psi_{ik}\sin\gamma_{jk} - v_j\cos\psi_{jk}\sin\gamma_{ik}}{\sin(\gamma_{jk} - \gamma_{ik})} \qquad (8)$$

$$\omega_k = \frac{-s_{ik}\cos\gamma_{jk} + s_{jk}\cos\gamma_{ik} - v_i\cos\psi_{ik}\cos\gamma_{jk} + v_j\cos\psi_{jk}\cos\gamma_{ik}}{d\sin(\gamma_{jk} - \gamma_{ik})}$$

$$\qquad (9)$$

The closed-loop linearized system is

$$\dot{l}_{ik} = k_1(l_{ik}^d - l_{ik}), \quad \dot{l}_{jk} = k_1(l_{jk}^d - l_{jk}), \quad \dot{\theta}_k = \omega_k. \qquad (10)$$

Separation Distance–To–Obstacle Control. This controller (denoted SD_oC) allows to avoid obstacles while following a leader. Thus, the outputs of interest are the separation l_{ij} between the follower and the leader, and the distance δ from an obstacle to the follower. We define a *virtual* robot R_o as shown in Figure 5.3, which moves on the obstacle's boundary with linear velocity v_o and orientation θ_o. For this case the

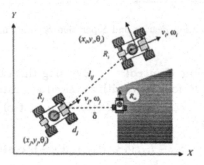

Figure 5.3. The Separation Distance to Obstacle Control $SDoC$.

velocity inputs for the follower robot R_j are given by

$$v_j = \frac{s_{ij}\cos\gamma_{oj} + s_{oj}\sin\gamma_{ij} + v_i\cos\psi_{ij}\cos\gamma_{oj}}{\cos(\gamma_{oj} - \gamma_{ij})} \qquad (11)$$

$$\omega_j = \frac{s_{ij}\sin\gamma_{oj} - s_{oj}\cos\gamma_{ij} + v_i\cos\psi_{ij}\sin\gamma_{oj}}{d\cos(\gamma_{oj} - \gamma_{ij})} \qquad (12)$$

Thus, the linearized kinematics become

$$\dot{l}_{ij} = k_1(l_{ij}^d - l_{ij}), \quad \dot{\delta} = k_o(\delta_o - \delta), \quad \dot{\theta}_j = \omega_j. \qquad (13)$$

where $s_{oj} \equiv k_o(\delta_o - \delta)$, δ_o is the desired distance from the robot R_j to an obstacle, and k_i's are positive controller gains.

It is worth noting that feedback I/O linearization is possible as long as $d\cos(\gamma_{oj} - \gamma_{ij}) \neq 0$, *i.e.*, the controller is not defined if $\gamma_{oj} - \gamma_{ij} = \pm k\frac{\pi}{2}$. This occurs when vectors $\vec{\delta}$ and $\vec{l}_{ij}$ are collinear.

By using this controller a follower robot will avoid the nearest obstacle within its *field-of-view* while keeping a desired distance from the leader. This is a reasonable assumption for many outdoor environments of practical interest. Complex environments (*e.g.*, star-like obstacles) are beyond the scope of this paper.

3.1.   A Basic formation building block

In this section we develop a general approach to build formations in a modular fashion. To be more specific, since each robot in the team is nonholonomic, it is able to control up two output variables [8], *i.e.*, a robot can follow another robot maintaining a desired separation and bearing, or follow two robots maintaining desired separations. Thus, a basic formation building block consists of a *lead* robot R_i, a *first follower* robot R_j, and a *follower* robot R_k. Figure 5.4 illustrates the basic formation and the actual robots we use in our experimental testbed. The basic idea is that R_i follows a given trajectory $g(t) \in SE(2)$, R_j and R_k use *SBC* and *SSC*, respectively.

Figure 5.4. The basic formation configuration.

3.2.   Stability analysis

In the following, we prove that the *basic formation* is stable, that is, relative distances and bearings reach their desired values asymptotically, and the internal dynamics of R_j and R_k are stable. Since we are using I/O feedback linearization [12], the linearized systems are given by (7) and (10) with outputs

$$z_1 = [l_{ij} \quad \psi_{ij}]^T, \qquad z_2 = [l_{ik} \quad l_{jk}]^T$$

It is straightforward to show that the output vectors $z_{1,2}$ will converge to the desired values arbitrarily fast. However, a complete stability analysis requires the study of the internal dynamics of the robots *i.e.*, the heading angles θ_j and θ_k which depend on the controlled angular velocities ω_j and ω_k.

Theorem 1 *Assume that the lead vehicle's linear velocity along the path* $g(t) \in SE(2)$ *is lower bounded i.e.*, $v_i \geq V_{\min} > 0$, *its angular velocity is also bounded i.e.*, $\|\omega_i\| < W_{\max}$, *the relative velocity* $\delta_v \equiv v_i - v_j$

and relative orientation $\delta_\theta \equiv \theta_i - \theta_j$ are bounded by small positive numbers ε_1, ε_2, and the initial relative orientations $\|\theta_i(t_0) - \theta_j(t_0)\| < c_1\pi$, $\|\theta_i(t_0) - \theta_k(t_0)\| < c_2\pi$ with $0 < c_{1,2} < 1$. If the control velocities (2)–(3) are applied to R_j, and the control velocities (8)–(9) are applied to R_k, then the formation is stable, and the system outputs l_{ij}, ψ_{ij}, l_{ik}, and l_{jk} converge exponentially to the desired values.

Proof: Let the system error $e = [e_1 \cdots e_6]^T$ be defined as

$$
\begin{aligned}
e_1 &= l_{ij}^d - l_{ij}, \quad e_2 = \psi_{ij}^d - \psi_{ij}, \quad e_3 = \theta_i - \theta_j \\
e_4 &= l_{ik}^d - l_{ik}, \quad e_5 = l_{jk}^d - l_{jk}, \quad e_6 = \theta_i - \theta_k
\end{aligned}
\tag{14}
$$

We need to show that the internal dynamics of R_j and R_k are stable which in formation control, is equivalent to show that the orientation errors e_3, e_6 are bounded. For the first follower R_j, we have

$$
\dot{e}_3 = \omega_i - \omega_j
$$

after some algebraic simplification, we obtain

$$
\dot{e}_3 = -\frac{v_i}{d}\sin e_3 + \eta_1(e_3, \omega_i, e_1, e_2)
\tag{15}
$$

where

$$
\eta_1(t, e_3) = (1 - \frac{l_{ij}}{d}\cos\gamma_{ij})\omega_i - \frac{1}{d}(k_1 e_1 \sin\gamma_{ij} + k_2 e_2 l_{ij}\cos\gamma_{ij})
$$

The nominal system *i.e.*, $\eta_1(t, e_3) = 0$ is given by

$$
\dot{e}_3 = -\frac{v_i}{d}\sin e_3
\tag{16}
$$

which is (locally) exponentially stable provided that the velocity of the lead robot $v_i > 0$. Since ω_i is bounded, it can be shown that $\|\eta_1(t, e_3)\| \leq \delta_1$. By using stability theory of perturbed systems [13], and the condition of the theorem $\|e_3(t_0)\| < c_1\pi$ for some positive constant $c_1 < 1$, then

$$
\|e_3(t)\| \leq \sigma_1, \qquad \forall\, t \geq t_1
$$

for some finite time t_1. Now for the follower R_k, the error system becomes

$$
\dot{e}_6 = \omega_i - \omega_k
$$

as before and after some work, we obtain

$$
\dot{e}_6 = -\frac{v_i}{d}\sin e_6 + \eta_2(e_6, \omega_i, e_4, e_5, \delta_v, \delta_\theta)
\tag{17}
$$

where

$$\eta_2(t, e_6) = \omega_i - \frac{v_i \delta_\theta \sin \psi_{jk} \cos(e_6 - \psi_{jk}) + \delta_v \cos(e_6 + \psi_{ik}) \cos \psi_{jk}}{d[\delta_\theta \cos(\psi_{jk} - \psi_{ij}) + \sin(\psi_{jk} - \psi_{ij})]} -$$
$$- \frac{k_5 e_5 \cos \gamma_{ij} + k_4 e_4 \cos \gamma_{jk}}{d[\delta_\theta \cos(\psi_{jk} - \psi_{ij}) + \sin(\psi_{jk} - \psi_{ij})]}$$

Again, the nominal system is given by (16) *i.e.*, $\eta_2(t, e_6) = 0$, and it is (locally) exponentially stable provided that the velocity of the lead robot $v_i > 0$. Since $\|\omega_i\| < W_{\max}$, $\|\delta_v\| < \varepsilon_1$, and $\|\delta_\theta\| < \varepsilon_2$, it can be shown that $\|\eta_2(t, e_6)\| \le \delta_2$. Knowing that $\|e_6(t_0)\| < c_2 \pi$ for some positive constant $c_2 < 1$, then

$$\|e_6(t)\| \le \sigma_2, \qquad \forall\, t \ge t_2$$

for some finite time t_2.

$\square$

The above theorem shows that, under some reasonable assumptions, the three-robot formation system is stable *i.e.*, there exists a Lyapunov function $V(t, e)$ in $[0, \infty) \times D$, where $D = \{e \in \Re^6 | \|e\| < c\}$, such that $\dot{V}(t, e) \le 0$. Let

$$V = e_{12}^T P_{12} e_{12} + \frac{1}{2} e_3^2 + e_{45}^T P_{45} e_{45} + \frac{1}{2} e_6^2 \qquad (18)$$

be a Lyapunov function for the system error (14) then

$$\dot{V} = -e_{12}^T Q_{12} e_{12} - e_{45}^T Q_{45} e_{45} - \frac{v_i}{d} e_3 \sin e_3 \qquad (19)$$
$$- \frac{v_i}{d} e_6 \sin e_6 + \eta_1(t, e_3) e_3 + \eta_2(t, e_6) e_6$$

where $e_{12}^T \equiv [e_1 \quad e_2]$, $e_{45}^T \equiv [e_4 \quad e_5]$, and P_{12}, P_{45}, Q_{12}, and Q_{45} are 2×2 positive definite matrices. By looking at (18)-(19), we can study some particular formations of practical interest.

- Let us assume two robots in a *linear motion* leader-following formation *i.e.*, v_i is constant, and $\omega_i = 0$. Thus the Lyapunov function and its derivative become

$$V_2 = e_{12}^T P_{12} e_{12} + \frac{1}{2} e_3^2 \qquad (20)$$

$$\dot{V}_2 = -e_{12}^T Q_{12} e_{12} - \frac{v_i}{d} e_3 \sin e_3 \qquad (21)$$

then the two-robot system is (locally) asymptotically stable *i.e.*, $e_3 \to 0$ as $t \to \infty$ provided that $v_i > 0$ and $\|e_3\| < \pi$. If ω_i is constant (circular motion), then e_3 is bounded. It is well-known that an optimal nonholonomic path can be planned by joining linear and circular trajectory segments. This result can be extended to n robots in a *convoy-like* formation (*c.f.*, [6]. Let us consider a team of n robots where R_i follows R_{i-1} under *SBC*. A Lyapunov function and its derivative can be given by

$$V_{1...n} = \sum_{i=2}^{n} e_{i-1,i}^T P_{i-1,i} e_{i-1,i} + \frac{1}{2} e_{\theta i}^2 \qquad (22)$$

$$\dot{V}_{1...n} = -\sum_{i=2}^{n} (e_{i-1,i}^T Q_{i-1,i} e_{i-1,i} + \frac{v_i}{d} e_{\theta i} \sin e_{\theta i} - \eta_i(t, e_{\theta i})) \qquad (23)$$

where $e_{i-1,i} = [l_{i-1,i}^d - l_{i-1,i} \quad \pi - \psi_{i-1,i}]^T$ is the output error, and $e_{\theta i} = \theta_{i-1} - \theta_i$ is the orientation error between R_{i-1} and R_i.

- A similar analysis can be carried out for the case of three robots in a *parallel* linear motion where $v_i = v_j = constant$, $\omega_i = \omega_j = 0$, and $\theta_i(t_0) = \theta_j(t_0)$. The Lyapunov function and its derivative are given by

$$V_3 = e_{45}^T P_{45} e_{45} + \frac{1}{2} e_6^2 \qquad (24)$$

$$\dot{V}_3 = -e_{45}^T Q_{45} e_{45} - \frac{v_i}{d} e_6 \sin e_6 \qquad (25)$$

then the three-robot system is (locally) asymptotically stable *i.e.*, $e_6 \to 0$ as $t \to \infty$ provided that $v_i > 0$, $\|e_6\| < \pi$ and $l_{ij} < l_{ik} + l_{jk}$. Again, this result can be extended to n robots in parallel linear formation.

3.3.   Coordination strategy

So far, we have shown that under certain assumptions a group of robots can navigate maintaining a stable formation. However, in real situations mobile robotic systems are subject to sensor, actuator and communication constraints, and have to operate within unstructured environments. These problems have motivated the development of a *switching paradigm* that allows robots change the shape of the formation *on-the-fly*. The basic strategy is as follows. Suppose a two-robot (R_1, R_2) formation is following a predefined trajectory using *SBC*. If there is an obstacle in the field-of-view of the follower, it switches to *SDoC*. When the obstacle has been successfully negotiated, R_2 switches back to *SBC*. Assume now a third robot R_3 joins the formation. Since R_3 has some sensor constraints, it may *see* or *follow* R_1, R_2 or both. For avoiding inter-robot collisions, the preferred configuration is that R_3 follows R_1 and R_2 using *SSC*. Thus, if R_3 sees only R_2, it will follow R_2 with desired values (*i.e.* l_{23}^d, ψ_{23}^d) selected in a way that R_3 is driven to the domain of controller *SSC*. Similarly, if R_3 sees only R_1, the desired output values (l_{13}^d, ψ_{13}^d) are chosen such that R_3 is driven to the domain of controller *SSC*. Furthermore, assume R_4 joints the group. It has six control possibilities to choose from as follows. R_4 may follow R_1, R_2, R_3, or any pair R_1R_2, R_1R_3, R_2R_3. The preferred configuration and desired values will depend on the prescribed formation shape and size.

This algorithm can be recursively extended to n robots. Let us consider a 4–robot case shown in Figure 5.5. To define a formation, we need

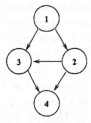

Figure 5.5. Control graph for 4 robots.

one separation–bearing control (R_2 following R_1) and two separation–separation controllers (R_3 following R_1 and R_2, and R_4 following R_2 and R_3). We call such a directed graph $\mathcal{H}$, a *control graph*. In a control graph, nodes and edges represent robots and control policies, respectively. Any control graph can be described by its *adjacency matrix* ([18]). For this example, the adjacency matrix becomes

$$H = \begin{bmatrix} 0 & 1 & 1 & 0 \\ 0 & 0 & 1 & 1 \\ 0 & 0 & 0 & 1 \\ 0 & 0 & 0 & 0 \end{bmatrix} \tag{26}$$

Note that this is a directed graph with the control flow from leader i to follower j. If a column k has a non zero entry in row i, then robot k is following i. A robot can have up to 2 leaders. The column with all zeros corresponds to the *lead* robot. A row with all zeros corresponds to a *terminal* follower.

It is clear that the number of possible control graphs increases dramatically with the number of robots. For labeled robots with the constraint of leaders having lower labels than followers, $n = 3$ allows 3 control graphs, $n = 4$ results in 18 graphs, and $n = 5$ results in 180 graphs.

The coordination strategy allows decentralized decision making for each individual robot. This is especially useful in simulations of large formations in complex scenarios to keep track of individual choice of controllers and switching between them. A formal study of control graphs in the context of formation control is a topic of current and future work [11].

4.   Trajectory generation using contact dynamics models

In this section, we propose a scheme for sensor-based trajectory generation. The key idea that distinguishes our approach from previous work is the use of rigid body contact dynamics models to allow collisions between the robot and its surroundings instead of avoiding them.

Consider a group of mobile robots moving in an environment with the presence of obstacles, we first characterize the surrounding spatial division of each mobile robot with three zones as depicted in Figure 5.6. Use robot R_1 as an example, the *sensing zone* denotes the region within which a robot can detect obstacles and other robots. The *contact zone* is a collision warning zone. The robot starts estimating the relative positions and velocities of any objects that may appear inside its contact zone. The innermost circle is the *protected zone* which is modeled as a rigid core during a possible contact. The *ellipse* within the protected zone represents the *reachable* region of the robot. It is pre-computed based on the robot's maximum kinetic energy. Thus, the actual robot is protected from collisions. In the planning process, we will use the protected zone as an abstraction of the robot itself.

The dynamics equations of motion for the ith nonholonomic robot in the group are given by

$$M_i(q_i)\ddot{q}_i + h_i(q_i,\dot{q}_i) \;=\; B_i(q_i)u_i - A_i(q_i)^T\lambda_i + \sum_{j=1}^{k}W_{ij}F_{ij} \quad (27)$$

$$A_i(q_i)\dot{q}_i \;=\; 0 \qquad\qquad\qquad (28)$$

where $q_i \in \Re^3$ is the vector of generalized coordinates, M_i is an 3×3 positive-definite symmetric inertia matrix, $h_i(q_i,\dot{q}_i)$ is a 3×1 vector of nonlinear inertial forces, B is a 3×2 input transformation matrix, A_i is a 3×1 matrix associated with the nonholonomic constraints, λ is the corresponding constraint force, and u_i is the 3×1 vector of applied (external) forces and torques. k is the number of the contacts between the ith–robot and all other objects which could be either obstacles or

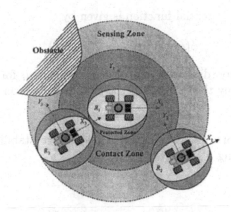

Figure 5.6. Zones for the computation of contact response.

other robots. $F_{ij} = (F_{N,ij} \ F_{T,ij})^T$ is a 2×1 vector of contact forces corresponding to the jth contact, and $W_{ij} \in \Re^{3\times2}$ is the Jacobian matrix that relates the velocity at the jth contact point to the time derivatives of the generalized coordinates of the robot.

We adopt a state-variable based compliant contact model described in [21] to compute the contact forces. At the jth contact of the agent i, the normal and tangential contact forces $F_{N,ij}$ and $F_{T,ij}$ are given by

$$F_{N,ij} = f_N(\delta_{N,ij}) + g_N(\delta_{N,ij}, \dot{\delta}_{N,ij}), \quad j=1,\ldots,k, \qquad (29)$$
$$F_{T,ij} = f_T(\delta_{T,ij}) + g_T(\delta_{T,ij}, \dot{\delta}_{T,ij}), \quad j=1,\ldots,k, \qquad (30)$$

where the functions f_N and f_T are the elastic stiffness terms and g_N and g_T are the damping terms in the normal and tangential directions respectively. Similar to handling rigid body contact, these functions can be designed to adjust the response of the robot. $\delta_{N,ij}(q)$ and $\delta_{T,ij}(q)$ are the local normal and tangential deformations which can be uniquely determined by the generalized coordinates of the system. The details and variations on the compliant contact model are discussed in [16, 21]. A key feature of this model is that it allows to resolve the ambiguous situations when more than three objects came into contact with one robot.

Figure 5.7 shows an example of an army-ant scenario in which 25 *holonomic* robots try to arrange themselves around a goal. The grouping is done dynamically using a decentralized decision making process. The team is initialized with two groups. A quadratic well type of potential function [14, 15] is constructed to drive the robots toward the goal. The expression of the potential function is given by

$$\phi(q) = \frac{k_p}{2}(q - q_g)^T(q - q_g) \tag{31}$$

where q_g is the coordinates of the goal. The input u_i for the ith agent can be obtained by the gradient of the potential function

$$u_i = -\nabla\phi(q_i) = -k_p(q - q_g) \tag{32}$$

which is a proportional control law. Asymptotic stabilization can be achieved by adding dissipative forces [14].

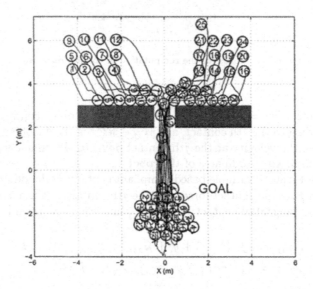

Figure 5.7. 25 holonomic robots arrange themselves around a target.

For the nonholonomic case substantial care in developing the local level controllers [3, 5] is required. We project the contact forces $\sum_{j=1}^{k} W_{ij}F_{ij}$ in (27) onto the reduced space while eliminating the constraint forces $A_i(q_i)^T\lambda_i$ in (28). As in [10], we use I/O linearization techniques to generate a control law u that stabilizes the robot's configuration about a reference trajectory, *i.e.*, $\|q - q_d\| \to 0$ as $t \to \infty$. The projected contact forces are treated as external disturbances during this process. We refer the reader to [11] for details.

This approach can be potentially scaled to tens and hundreds of robots. Each vehicle is driven by a set of local controllers that uses the information obtained within its local sensing zone. Thus, explicit inter–robot communication is avoided. We will illustrate the application of this method in the next section.

5. Simulation results

We illustrate our approach by an example in which 4 nonholonomic mobile robots $R_{1,2,3,4}$ are commanded to a goal position within an unknown environment. In the first experiment, the robots are autonomous, and the formation constraint is not explicitly enforced. Each robot runs its own trajectory generator and controller. As it can be seen in Figure 5.8, the robots are able to navigate and reach the goal position.

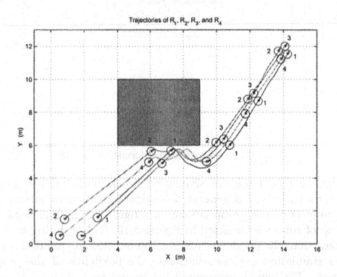

Figure 5.8. Decentralized trajectory generation and control.

In a second experiment, the trajectory generator produces a trajectory only for the lead robot R_1. Then, the basic controllers and the coordination strategy outlined in section 3 are implemented on $R_{2,3,4}$. The desired shape of the formation is a *diamond* with inter-robot separation of 1.2 m. In order to reach the desired formation, R_2 follows R_1 with $SB_{12}C$. R_3 has to maintain a specified distance from R_1 and R_2, *i.e.*, $S_{13}S_{23}C$. Similarly, R_4 is to maintain a specified distance from R_2 and R_3, *i.e.*, $S_{24}S_{34}C$. However, $R_{2,3,4}$ may switch controllers depending on their relative positions and orientations with respect to neighboring robots or obstacles. Thus, for a initial formation, the group is able to reconfigure itself until the desired formation shape is achieved. Figure

5.9 shows that robots are able to negotiate the obstacle, avoid collisions and keep the formation shape.

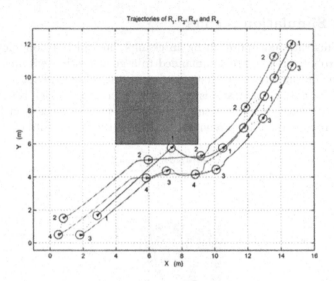

Figure 5.9. Leader–following formation control.

Finally, we repeat the last simulation experiment, but this time the lead robot's trajectory is generated considering the mass/inertia of the entire group (*i.e.*, the group is seen as a rigid body). The behavior of the group of robots is depicted in Figure 5.10. The resulting trajectory is different than the previous case, but computationally more intensive.

These simulation exercises illustrate the flexibility of the proposed framework. We have implemented behavior–based navigation, leader-following and virtual structure formation in a straightforward manner. This flexibility will be useful in actual multi-robotic missions where autonomous vehicles may need to navigate in a decentralized fashion until a target has been detected. Then the group can switch to a leader-following behavior and keep a desired formation shape around the target. Finally, a rigid formation (desired shape and size) behavior can be switched in to manipulate and transport the target.

6.   Conclusions

In this paper, we presented a framework for controlling and coordinating a group of nonholonomic mobile robots. The framework integrates three key components for cooperative control of robot formations: (1) reference trajectory generation, (2) a coordination strategy that allows

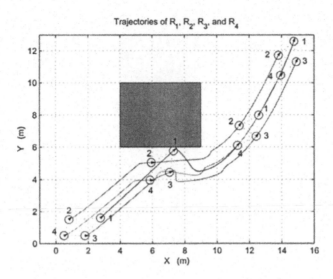

Figure 5.10.   Centralized trajectory generation with decentralized control.

the robots to switch between control policies, and (3) a suite of controllers that under reasonable assumptions guarantees stable formations. Our approach can easily scale to any number (tens and hundreds) of vehicles and is flexible enough to support many formation shapes. The framework described here can also be applied to other types of unmanned vehicles (*e.g.*, aircraft, spacecraft, and underwater vehicles). Currently, we are formalizing and extending the coordination strategy for a large number of robots, and conducting experiments on a group of car-like mobile platforms equipped with on-board omni–directional vision system. Also, we are applying these ideas to formation flight of multiple unmanned aerial vehicles *UAVs* on $SE(3)$

Acknowledgments

This research was supported in part by DARPA ITO MARS 130-1303-4-534328-xxxx-2000-0000, DARPA ITO MOBIES F33615-00-C-1707, and NSF grant CISE/CDS-9703220.

References

[1] Alur, R., Das, A., Esposito, J., Fierro, R., Hur, Y., Grudic, G., Kumar, V., Lee, I., Ostrowski, J. P., Pappas, G., Southall, J., Spletzer, J., and Taylor, C. J. (2000). A framework and architecture for multirobot coordination. In *Proc. ISER00, Seventh International Symposium on Experimental Robotics*, Honolulu, Hawaii.

[2] Barraquand, J., Langlois, B., and Latombe, J. (1992). Numerical potential field techniques for robot path planning. *IEEE Trans. Syst., Man, and Cyber.*, 22(2):224–241.

[3] Barraquand, J. and Latombe, J. (1993). Non-holonomic multibody mobile robots: controllability and motion planning in the presence of obstacles. *Algorithmica*, 10:121–155.

[4] Beard, R. W., Lawton, J., and Hadaegh, F. Y. (1999). A coordination architecture for spacecraft formation control. Submitted to IEEE Trans. Contr. Sys. Technology.

[5] Bemporad, A., De Luca, A., and Oriolo, G. (1996). Local incremental planning for a car-like robot navigating among obstacles. In *Proc. IEEE Int. Conf. Robot. Automat.*, pages 1205–1211.

[6] Canudas-de-Wit, C. and NDoudi-Likoho, A. D. (2000). Nonlinear control for a convoy-like vehicle. *Automatica*, 36:457–462.

[7] Das, A., Fierro, R., Kumar, V., Southall, J., Spletzer, J., and Taylor, C. J. (2001). Real-time vision based control of a nonholonomic mobile robot. To appear in IEEE Int. Conf. Robot. Automat., ICRA01.

[8] De Luca, A., Oriolo, G., and Samson, C. (1998). Feedback control of a nonholonomic car-like robot. In Laumond, J.-P., editor, *Robot Motion Planning and Control*, pages 171–253. Springer-Verlag, London.

[9] Desai, J., Ostrowski, J. P., and Kumar, V. (1998). Controlling formations of multiple mobile robots. In *Proc. IEEE Int. Conf. Robot. Automat.*, pages 2864–2869, Leuven, Belgium.

[10] Fierro, R. and Lewis, F. L. (1999). Robot kinematics. In Webster, J., editor, *Wiley Encyclopedia of Electrical and Electronics Engineering.* John Wiley and Sons, Inc.

[11] Fierro, R., Song, P., Das, A., and Kumar, V. (2001). A framework for scalable cooperative navigation of autonomous vehicles. Technical report MS-CIS-01-09, Department of Computer and Information Science, University of Pennsylvania, Philadelphia PA, USA.

[12] Isidori, A. (1995). *Nonlinear Control Systems.* Springer-Verlag, London, 3rd edition.

[13] Khalil, H. (1996). *Nonlinear Systems.* Prentice Hall, Upper Sadle River, NJ, 2nd edition.

[14] Khatib, O. (1986). Real-time obstacle avoidance for manipulators and mobile robots. *International Journal of Robotics Research*, 5:90–98.

[15] Koditschek, D. (1987). Exact robot navigation by means of potential functions: Some topological considerations. In *Proc. IEEE Int. Conf. Robot. Automat.*, pages 1–6.

[16] Kraus, P. R. and Kumar, V. (1997). Compliant contact models for rigid body collisions. In *Proceedings of IEEE International Conference on Robotics and Automation*, pages 1382–1387.

[17] Liberzon, D. and Morse, A. S. (1999). Basic problems in stability and design of switched systems. *IEEE Control Systems*, 19(5):59–70.

[18] Nemhauser, G. L. and Wolsely, L. A. (1988). *Integer and Combinatorial Optimization*, chapter I.3. Wiley.

[19] Parker, L. E. (2000). Current state of the art in distributed autonomous mobile robotics. In Parker, L. E., Bekey, G., and Barhen, J., editors, *Distributed Autonomous Robotic Systems*, volume 4, pages 3–12. Springer, Tokio.

[20] Rimon, E. and Koditschek, D. (1992). Exact robot navigation using artificial potential fields. *IEEE Trans. Robot. & Autom.*, 8(5):501–518.

[21] Song, P., Kraus, P., Kumar, V., and Dupont, P. (2001). Analysis of rigid–body dynamic models for simulation of systems with frictional contacts. *Tran. ASME*, 68:118–128.

[22] Spletzer, J., Das, A., Fierro, R., Taylor, C. J., Kumar, V., and Ostrowski, J. P. (2001). Cooperative localization and control for multi-robot manipulation. Submitted to IEEE/RSJ Int. Conf. Intell. Robots and Syst., IROS2001.

[23] Volpe, R. and Khosla, P. (1990). Manipulator control with superquadric artificial potential functions: Theory and experiments. *IEEE Trans. on Syst., Man, and Cyber.*, 20(6):1423–1436.

Chapter 6

COOPERATIVE BEHAVIOR SCHEMES FOR IMPROVING THE EFFECTIVENESS OF AUTONOMOUS WIDE AREA SEARCH MUNITIONS*

Daniel P. Gillen
Student, Air Force Institute of Technology, AFIT/ENY
Wright Patterson Air Force Base, OH
Daniel.Gillen@afit.af.mil

David R. Jacques
Assistant Professor, Air Force Institute of Technology, AFIT/ENY
Wright Patterson Air Force Base, OH
David.Jacques@afit.af.mil

Abstract

The problem being addressed is how to best find and engage an unknown number of targets in unknown locations (some moving) using multiple autonomous wide area search munitions. In this research cooperative behavior is being investigated to improve the overall mission effectiveness. A computer simulation was used to emulate the behavior of autonomous wide area search munitions and measure their overall expected performance. This code was modified to incorporate the capability for cooperative engagement based on a parameterized decision rule. Using Design of Experiments (DOE) and Response Surface Methodologies (RSM), the simulation was run to achieve optimal decision rule parameters for given scenarios and to determine the sensitivities of those parameters to the precision of the Autonomous Target

---

*The views expressed in this article are those of the authors and do not reflect the official policy or position of the United States Air Force, Department of Defense, or the US Government.

R. Murphey and P.M. Pardalos (eds.), Cooperative Control and Optimization, 95–120.
© 2002 *Kluwer Academic Publishers.*

Recognition (ATR) algorithm, lethality and guidance precision of the warhead, and the characteristics of the battlefield.

Keywords: cooperative engagement, cooperative behavior, autonomous munitions, wide area search munitions.

1. Introduction

1.1. General

The problem being addressed is how to best find and engage an unknown number of targets in unknown locations (some moving) using multiple cooperating autonomous wide area search munitions. The problem is exacerbated by the fact that not all target priorities are the same, the munition target discrimination capability is never perfect, and target destruction is never a certainty even once engaged. Further, factors such as clutter density throughout the battlefield and ratio of targets to civilian or military non-targets create even more complications for these smart, yet simple-minded, munitions.

This research does not necessarily provide the precise solution to this rather complex problem; rather, this research provides a possible methodology for how to attack this problem using different optimization methodologies and shows some sample results.

This research was sponsored by the Munitions Directorate of Air Force Research Labs at Eglin Air Force Base (AFB). All research took place at the Air Force Institute of Technology (AFIT), Wright-Patterson AFB, Ohio.

1.2. Background

The United States Air Force has significantly reduced the size of its military forces as a response to changing national military objectives and diminishing budgets. This reality has forced the Air Force to look for more cost effective ways of achieving its extremely crucial mission. One development has been the creation of small, lightweight, low-cost, autonomous munitions fully equipped with INS/GPS navigation and seekers capable of Autonomous Target Recognition (ATR). The intent in using these autonomous munitions is to employ larger numbers of cheaper, less sophisticated munitions as opposed to fewer numbers of expensive, complex munitions. However, in order to realize the full capabilities of a system composed of large numbers of smaller subsystems (or agents), the individual agents must behave cooperatively. Methods of evaluating mission effectiveness of these munitions have previously been developed for the case of non-cooperating munitions [7]. In this research cooperative behavior is being investigated to improve the overall mission effectiveness.

In a study provided by RAND [6], the rationale was developed for investigating cooperative behavior between Proliferated Autonomous Weapons (PRAWNs). They showed by implementing a cooperative weapon behavior logic into a computer simulation that there was a definite added potential when cooperation was incorporated into the logic of PRAWNs. This study supported the hypothesis that while the individual munitions may be less capable than conventional munitions under development today, through communication across the swarm of weapons, the group exhibits behaviors and capabilities that can exceed those of more conventional systems that do not employ communication between weapons. The potential benefits which come about through shared knowledge include relaxed sensor performance requirements, robustness to increases in target location errors, and adaptivity to attrition and poor target characterization.

In this study, however, a fixed decision rule (called "swarming algorithm") was used. This algorithm was based on the foundations of two areas of study: ethology (the science of animal behavior) and robotics developed in the civil sector. The collective intelligence that seems to emerge from what are often large groups of relatively simple agents is what the engineers of the RAND study tried to capture in their swarming algorithm. While this algorithm worked for what they were doing, the research did not show how this decision algorithm compared to other possible decision algorithms. Also, the RAND study concentrated on a very specific battlefield layout that was composed of large clusters of targets and no possibility of encounters with non-targets or clutter. By not taking into account non-targets or clutter, the munitions had no false target attacks. According to Jacques [7], methods and models for evaluating the effectiveness of wide area search munitions must take into account the degradation due to false target attacks.

Scientists studying animal behavior have identified and analytically modeled many behaviors of natural organisms that have parallels to the tasks that weapons must achieve in order to search for, acquire, and attack targets. Some of these studies include Reynolds' considerations for the formation of flocks, herds, and schools in simulations in which multiple autonomous agents were repulsed by one another (and other foreign objects) by inverse square law forces [12] and Dorigo's studies of ant colony optimizations [5]. Scientists in the field of robotics have developed architectures for the controlling of individual robots or agents, which allow groups of individuals to experience the benefits of group or swarm behaviors. These include the studies by Arkin, Kube and Zhang, Asama, and Kwok. Arkin demonstrated an approach to cooperation among multiple mobile robots without communications [1], and Kube and Zhang also researched the use of decentralized robots performing various tasks without explicit communication [8]. Asama sums up the challenge in choosing the right behaviors for your agents by saying that "an autonomous and decentralized system has two essentially contradictory characteristics, autonomy and cooperativeness, and the biggest problem in the study of distibuted autonomous robotic systems is how to reconcile these two features" [2]. Kwok considered the problem of causing multiple (100's) of autonomous mobile robots to converge to a target using an on-board, limited range sensor for sensing the target and a larger but also limited-range robot-to-robot communication capability [9].

While much of the research in the field of cooperative control of robotics has been able to apply some of the basic principals learned from ethology, the application to cooperative engagement of autonomous weapons is rather limited. Since each of the munitions has a specific Field of View (FOV) on the order of a half mile in width, the munitions are normally programmed to fly a half mile from each other in order to limit the FOV overlap. Scenarios exist where large FOV overlap is desired in the interest of redundant coverage and higher probabilities of success, but the study of these scenarios is more applicable to the cooperative search problem than the cooperative engagement problem. Therefore, the protection and drag efficiencies gained by flocking, schooling or herding are not applicable to this study. However, the concept of ant foraging does have application to the problem at hand. Moreover, what if the ants had the ability to *choose* to follow the pheromone deposits to the known source of food or to *choose* to seek out a different area for a possible larger, better, or closer food source? By what criteria could this decision be made? Is the decision criteria the same for all situations? Taking this analogy one step further (and maybe a little beyond reality), what happens when an ant falsely identifies a poisonous food source as a good food source and causes the colony to subsist off of this unknown danger? These questions have not been answered in the applied research of robotics but are extremely important questions for the application of cooperative control of autonomous wide area search munitions.

1.3. Objectives

The primary objective of this study was to investigate the use of cooperative behavior to improve the overall mission effectiveness of autonomous wide area search munitions. The specific objectives were to:

1 Establish a methodology for measuring the expected effectiveness of a cooperative system of wide area search munitions

2 Develop optimal cooperative engagement decision rules for a variety of realistic scenarios

3 Qualitatively analyze the sensitivities of the decision rule parameters to the precision of the submunition's ATR algorithm, the lethality and guidance precision of the warhead, and the characteristics of the battlefield (clutter density, target layout, etc.)

2. Baseline computer simulation

This Monte Carlo based Fortran program was originally developed by Lockheed Martin Vought Systems [10] as an effectiveness model for the Low Cost Autonomous Attack System (LOCAAS). However, it is versatile enough to be used for any generic wide area search munition. The simulation makes no attempt to model the aerodynamics, guidance, etc. of the submunitions, however, it does model multiple submunitions in a coordinated search for multiple targets. Prior to the modifications made through this research, this program had the capability to simulate the following events of the submunition "life cycle":

- Round dispense (any number of rounds)

- Submunition dispense (any number of submunitions per round)

- Submunition flies a user supplied pattern by following predetermined waypoints and looks for targets on the ground

- If a target enters a submunition's FOV, the submunition may acquire it based on the probabilities associated with the ATR algorithm

- Once acquired, the submunition can select that target to engage

- Once engaged, the submunition attempts to hit the target

- Once the target is hit, an assessment is made as to whether the target has been completely destroyed (dead) or is still in working condition (alive)

The simulation allows for any number of targets with varying priority levels, the addition of non-targets (military or civilian), and a user supplied clutter density per square kilometer of battlefield. The simulation is extremely flexible in its capabilities to handle a multitude of input parameters and supplies all sorts of results as output files at the conclusion of each run.

2.1.   Inputs to the simulation

To run the simulation, two separate input files are required. The first contains the information concerning the user supplied flight paths for the submunitions once dispensed from the rounds including waypoints, altitude and velocity, and the second contains all the parameters characterizing the submunitions and the parameters required to run the simulation. Table 6.1 shows a summary of some of the input parameters that must be entered regarding the characteristics of the submunitions and targets.

Since the munition effectiveness is determined by the outcome of Monte Carlo runs, the user also has the ability to pick a baseline seed (which is modified for every repetition in a series) and the number of Monte Carlo trials.

2.2.   Outputs of the simulation

The main output file for the simulation lists all of the input parameters used to run the simulation for tracking purposes. Then for each Monte Carlo repetition, a brief history of what each submunition did during that repetition is displayed. Finally, at the end of the main output file, all Monte Carlo repetitions are summarized showing a breakdown, per target type and per individual target, of the number of acquisitions, selections, hits, and kills, as well as the total number of kills and unique kills for that simulation run.

3.   Simulation modifications

The baseline simulation has some capacity for cooperative engagement, but it was insufficient for purposes of this research. Specifically, cooperative attack decisions have to be made immediately upon target declaration, the number of submunitions to be redirected is set a priori, and there is no provision for expanded search if the target is not found by the submunition being redirected. For these reasons, significant modifications were made to the simulation.

Table 6.1. Input Parameters to Baseline Simulation

Parameter	Description
Numbers:	
Rounds	Total number of rounds dispensed
Submunitions	Total number of submunitions (and submunitions per round)
Target Types	Priority 1, priority 2, non-targets, etc.
Targets	Total number of targets and how many of each type
Discrete:	
Random Targets	Either targets are placed in specific locations or random within a specified area ≤ total battlefield area
Blind in Turns	Submunition's target detection is turned off when turning
Reliabilities:	
Round	Probability that round will not fail
Submunition	Probability that submunition will not fail
Probabilities:	(Input for each target type)
Acquisition	Submunition will acquire the target when it enters its FOV
Hit	Submunition will hit the target once its acquired
Kill	Submunition will kill the target once its hit
Correct Identification	Submunition will identify the target correctly or incorrectly (incorrect identifications are distributed among all target types as desired)
Seeker Data:	
Foot Print Width	Width of the FOV on the ground
Beam Width	Beam width in degrees used for vertical FOV
Boresight Angle	Angle at which the LADAR points down from the horizon
Scan Time	Time for the FOV to sweep the entire foot print width
Flyback Time	Time for the FOV to return at the completion of each sweep
Submunition Data:	
Min Turn Radius	Minimum turn radius the submunition can fly
Time of Flight	Total Time of flight from submunition dispense time to expiration
Target Data:	
Locations	Specific locations of all targets if using non-random target layout
Mobility Data	If mobile: start time, heading, speed, acceleration time

3.1.   Redirecting submunitions

The first step in the modification process was to be able to redirect any number of submunitions at any time toward any found targets. The way this was accomplished was using a structured array to store all target information on the targets found. This structured array stored the x, y and z coordinates of the target (if the target was a moving target, these coordinates would be those corresponding to the position of the target at the time it was acquired and selected), and the type of target found. A very important distinction which needs to be made at this point is that the target type stored is not necessarily the correct identification of the target found; it is the identification of the target type determined by the munition that identified that target. Therefore, the type of target found which is stored in this target array may not be the true type of target located at the stored coordinates.

Once the target information was stored, a method for distributing that information had to be determined. Obviously, since this was just a simulation, it would be easy to just provide all submunitions access to all entries in the aforementioned structured array. But is this feasible, realistic or even advantageous? Since this study hoped to gain some insight into the trade-offs between local and global communication, a mechanism for determining whether a submunition received the communicated information had to be implemented. First of all, in this study incomplete communications were not considered, i.e., either a submunition receives all the information about the target or none. However, communications reliability based solely on whether or not the submunition was within a certain maximum communications range didn't seem too realistic either. Therefore, a communications reliability function was developed. In order to keep it relatively simple, this function was solely based on the probability of communication failures increasing as maximum communications range was approached. Maximum communication is not set a priori; rather it is one of the variables to be determined by the design optimization process. The reliability function used is shown in equation (1).

Comm Rel $=$

$$
\begin{cases}
1 & \text{if range} \leq \frac{\text{max comm range}}{2} \\
\left(\frac{\text{max comm range} - \text{range}}{\frac{\text{max comm range}}{2}}\right)^{0.1} & \text{if range} > \frac{\text{max comm range}}{2}
\end{cases}
\tag{1}
$$

Figure 6.1 illustrates an example of this communications reliability function for a maximum communications range of 10,000 meters.

In order to implement the decision algorithm described in the following section, the amount of information that had to be shared among the submunitions had to be determined. For practical implementation concerns, there was a desire to limit the amount of communication required and to limit all communication to words or numbers as opposed to images. This low bandwidth communication seemed most feasible for

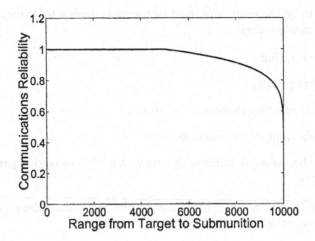

Figure 6.1. Communications Reliability Function

this application. For this study, the following three pieces of information were communicated for each target found:

- Location of the target

- Type of target

- Specific target to be engaged

The location of the target was communicated as the precise x and y coordinates of the target. The type of target was communicated as either a high priority (priority 1) or a low priority (priority 2). The specific target to be engaged is, in reality, a very difficult piece of information to communicate and keep track of reliably, especially with non-global and non-perfect communications. However, in this study, the target registration problem was not considered.

3.2. Decision algorithm

The purpose of the decision algorithm was to provide a criteria by which the submunitions could "decide" whether or not to participate in a cooperative engagement. In developing the algorithm, the goals were to incorporate all important factors that should be taken into account for making a cooperative engagement decision and to keep it simple since the available computing power aboard these submunitions is minimal. After several iterations, the following (in no particular order) were determined to be the most important factors that needed to be included in the decision algorithm:

- Fuel remaining

- Target priority

- Range rate from submunition to target

- Range from submunition to target

- Number of submunitions that have already engaged a particular target

To keep the decision algorithm simple, the basic first order expression shown in equation (2) was used.

$$\text{Threshold} = \alpha_1 * x_1 + \alpha_2 * x_2 + \alpha_3 * x_3 - \alpha_4 * x_4. \qquad (2)$$

where

$$
\begin{aligned}
x_1 &= \text{Normalized Fuel Remaining} \\
x_2 &= \text{Normalized Target Priority} \\
x_3 &= \text{Normalized Range Rate} \\
x_4 &= \text{Normalized Number of Engaged Submunitions on a} \qquad (3) \\
&\quad \text{particular target} \\
\alpha_i &= \text{Weighting Parameters}
\end{aligned}
$$

Normalizing the fuel remaining in the simulation was easily accomplished by normalizing time of flight or search time. Since each submunition had a twenty minute total search time, the normalized time of flight was the current time divided by 1200 seconds. Target priority was normalized by assigning a value of one to a priority one target, one-half to a priority 2 target and zero for anything else.

The purpose of incorporating a range rate parameter in the decision rule was to apply little influence on the decision (or even discourage a cooperative engagement) when the range rate was negative (the submunition is moving towards the target) and to encourage a cooperative engagement when the range rate is positive (the submunition is moving away from the target). This provided a means for allowing the submunition to continue its predetermined search pattern if it was flying in the general direction toward a known target location. The expression used to normalize range rate is shown in equation (4) with $\dot{r}$ defined by a backward difference.

$$\text{normalized range rate} = \left| \frac{\dot{r} - \text{vel}}{2 * \text{vel}} \right| \tag{4}$$

where

$$\dot{r} = \frac{\text{range}_i - \text{range}_{i-1}}{\text{time}_i - \text{time}_{i-1}}$$

Figure 6.2 illustrates the function for normalized range rate shown in equation (4).

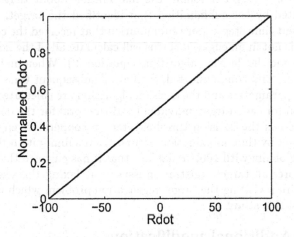

Figure 6.2.   Normalized Range Rate

The actual range from the submunition to a specific target is not explicitly used in this decision algorithm, however, a range check was added to the simulation to ensure that a submunition is not redirected toward a target that cannot be reached based on insufficient fuel remaining.

The last parameter in the decision algorithm is the normalized number of engaged submunitions on a specific target. The purpose of this parameter is to discourage multiple cooperative engagements on a single target in an attempt to spread out the total hits and not send all submunitions after the same target. When a target has been engaged by only a single submunition, then this parameter should not be discouraging a cooperative engagament on that target. However, once one submunition has cooperatively engaged a target (resulting in a total of two munitions attempting to hit that specific target), this parameter should be invoked to discourage any additional submunitions from choosing to cooperatively engage that target. Equation (5) was used to normalize this parameter.

Normalized Parameter = Number of engaged submunitions − 1   (5)

Note that in equation (2) a "-" sign is implemented in front of this parameter in order to *discourage* a cooperative engagement as this parameter increases. As desired, this parameter equals zero when only one submunition has engaged a specific target but then increases in value as more submunitions cooperatively engage that target.

The implementation of the decision rule in equation (2) was rather simple. Once a target is found, the information about that target is communicated by the submunition that identified the target. Then at all subsequent time steps, every submunition that received the communication and is not in an engaged status will calculate all of the normalized parameters and the decision algorithm, equation (2). When multiple targets are found and communicated, then at all subsequent time steps the normalized parameters and the decision algorithm are calculated by each submunition for each target individually. If the total for the decision algorithm exceeds the decision threshold, then a cooperative engagement on that target by that submunition occurs. That submunition then communicates globally with 100% reliability that it has engaged that specific target (ignores all target registration issues). Relaxing this assumption would require revisiting the target registration problem which is beyond the scope of this study.

3.3.   Additional modifications

In order to best achieve the objectives of this study, a few additional changes needed to be made to the simulation. The first was a simple modification to the main output file to include the values of each normalized parameter as well as the weights on the parameters every time a cooperative engagement was invoked. This provided a means to track all cooperative engagements, and ensure the decision algorithm was being implemented properly.

A second change was an attempt to answer the following question: what should happen to a submunition that is sent off its original search pattern to cooperatively engage a target that it cannot find? This situation could result from a failure of the ATR algorithm on either the original munition that identified the "target", or the munition searching for the previously found target. In order to accomodate this, an attempt was made to create a new search pattern for the redirected submunition that focused on the location of the target that was cooperatively engaged. The new pattern used was a growing figure-8 centered on the communicated target location. This pattern would initially turn the submunition around after it crossed the expected target location to fly right back over it as an attempt to acquire and classify the target if it simply "missed" it the first time. If the target was still not selected on this second pass, then the submunition would continue flying past the target, but this time farther past the target, in the opposite direction in which the submunition first approached the target area. It would then turn around and fly back toward the target until the submunition engages a target or expires (search time depletes)–the submunition cannot participate in a second cooperative engagement on a different target.

The behavior chosen to handle this situation is not necessarily that which would be implemented operationally, nor was any research completed that showed this behavior would produce optimal results. This situation was deemed outside the scope of the research and could better be addressed by a study in cooperative search.

4.   Applied response surface methodologies

Response surface methodology is a collection of statistical and mathematical techniques useful for developing, improving, and optimizing processes. Most applications include situations where several input variables potentially influence some performance measure or quality characteristic of the system. The purpose of RSM is to approximate the measures of performance, referred to as response functions, in terms of the most critical factors (or independent variables) that influence those responses. In doing this, a response surface can then be mapped out showing how variations in the independent variables affect the responses.

A typical RSM according to Myers and Montgomery [11] is broken into phases. The first experiment is usually designed to investigate which factors are most influential to the responses with the intent of eliminating the unimportant ones. This type of experiment is usually called a screening experiment and is typically referred to as phase zero of a response surface study. Once the important independent variables are identified, phase one begins with the intent of determining if the current levels or settings of the independent variables result in values of the responses near optimum. If the current settings or levels of the independent variables are not consistent with optimum performance, then adjustments to the input variables that will move the responses toward the optimum must be made. To do this a first order model and the optimization technique called the method of steepest ascent are employed. Phase two of the response surface study begins when the process is near the optimum. At this point models that will accurately approximate the true response functions are desired. Because the true response surfaces usually exhibits curvature near the optimum, models of at least second order must be used. Then, finally, the models of the various responses must be analyzed to determine the settings for the independent variables that provide the optimal expected performance over all responses.

For this study, RSM was used to determine the optimal settings for the α_i's in the cooperative engagement decision algorithm shown in equation (2) as well as the maximum communications range. The optimal α_i's were simply the weights on the parameters used in the decision rules. A high weight on a parameter means that that parameter is of high importance in making the decision to cooperatively engage, whereas a low weight on a parameter can be interpreted to mean that that parameter is less important (or even insignificant) in making the decision to cooperatively engage. A low maximum communications range implies that local communications are employed, whereas a high maximum communications range implies global communications.

4.1.   Independent variables

Because of the relatively small number of input variables, the phase zero screening experiments were not necessary for this study. Therefore, the first step was to choose the ranges of the independent variables to begin the RSM. A decision threshold for equation (2) was, without loss of generality, chosen equal to one (the α_i's could be scaled to accommodate a non-unity threshold value). In picking the ranges for the independent variables, careful consideration was made to ensure the RSM studies would be investigating the effects of different cooperative engagement decision rules. Therefore, the values for the independent variables when chosen at their extremes had to be able to result in the possible triggering of the decision algorithm. Since the first three parameters in the decision rule (as described on page 104) were normalized to have maximum values of one, the weights on these parameters could not be all less than one-third or else a cooperative engagament would be impossible. This would, therefore, result in an RSM study investigating the effects of different cooperative engagament decision rules *and* no cooperation at all. Since this was not the goal, minimum values for the first three parameters had to be chosen greater than one third. Table 6.2 shows the values used for each independent variable.

The values chosen in Table 6.2 ensure that even when the independent variables are chosen at their extremes, cooperative engagements are still possible. The maximum communications range values were chosen based on a battlefield that was approximately 300 square kilometers in size.

Table 6.2. Independent Variable Ranges for RSM

Variable	Weight on	Minimum	Maximum
α_1	Time of Flight	0.4	0.8
α_2	Target Priority	0.4	0.8
α_3	Range Rate	0.4	0.8
α_4	Number of Munitions	0.4	0.8
Maximum Communications Range		5 km	15 km

4.2.   Responses

The responses had to be chosen to somehow accurately measure the expected mission effectiveness for wide area search munitions. Four responses were chosen:

- Unique Kills

- Total Kills

- Total Hits

- Target Formula

Unique kills was defined as the expected number of unique, real targets killed (each target can only be killed once). Total kills was defined as the expected number of submunitions to put lethal hits on a real target. Total hits was defined as the expected number of real target hits. Finally, the target formula response was used as a means of measuring the hits on high priority targets (priority one) versus hits on low priority targets (priority two) and incorporating a penalty for any hits on non-targets. This target formula is shown in equation (6) where "# prior 1 hits" means the number of hits on priority one targets, "# prior 2 hits" means the number of hits on priority two targets, and "# non-target hits" means the number of hits on any non-targets.

$$\text{Tgt Formula} = 2*(\text{\# prior 1 hits}) + (\text{\# prior 2 hits}) - (\text{\# non-target hits}) \tag{6}$$

A simple example can be used to distinguish and better understand these responses. This example has five submunitions, 2 real targets (one high priority and one low priority) and one non-target. Submunition #1 hits target #1, a high priority target, but does not kill it. Submunition #2 hits that same target (target #1) but kills it. Submunition #3 hits and kills target #2, a low priority target. Then submunition #4 also hits target #2, and this engagement is also deemed a kill (even though the target was already dead) . Finally, submunition #5 hits the non-target. The responses for this example are shown in Table 6.3.

Table 6.3. Responses for Example

Response	Explanation	Value
Unique Kills	targets #1 and #2	2
Total Kills	submunitions #2, #3, and #4	3
Total Hits	submunitions #1, #2, #3, and #4	4
Target Formula	2 hits on target #1 (high priority target)	
	2 hits on target #2 (low priority target)	
	1 hit on a non-target	5

4.3.   Phase 1

The purpose of phase one is to determine if the current ranges for the independent variables shown in Table 6.2 result in responses that are near optimal. To accomplish this a 2^{5-1} fractional factorial design was used. This orthogonal resolution V design required a total of 16 runs to complete. Each design was augmented with four center runs resulting in a total of 20 runs. For each run the values for each of the four responses were recorded. Using an analysis of variance (ANOVA) for each response, an attempt to fit first order models to each response was made. Whenever a first order model was appropriate, the method of steepest ascent was used to traverse the response surface to a new operating region that was closer to the optimal design point. The method of steepest ascent is summarized by the following few steps.

1 Fit a planar (first-order) model using an orthogonal design

2 Compute a path of steepest ascent where the movement in each design variable direction is proportional to the magnitude of the regression coefficient corresponding to that design variable with the direction taken being the sign of the coefficient

3 Conduct experimental runs along the path

4 Choose a new design location where an approximation of the maximum response is located on the path

5 Conduct a second fractional factorial experiment and attempt to fit another first order model

If a second first order model is accurately fit, then a second path of steepest ascent can be computed and traversed until a region is reached where a higher-order model is required to accurately predict system behavior.

In this study, after the initial fractional factorial design was completed, a first order model was never adequate. Therefore, the method of steepest ascent was never required because the starting region of design seemed to always be close enough to the optimal point over all responses.

4.4.   Phase 2

The purpose of this phase is to build second (or higher) order models to accurately predict all responses and choose the settings for the independent variables that will result in the optimal expected performance over all responses. Since the resolution V fractional factorial was already completed at the appropriate design point, the ideal second order design would be able to simply augment the first design to decrease the total number of runs, thereby saving time and money. Therefore, the Central Composite Design (CCD) was used. This design requires three parts:

- Two-level factorial design or resolution V fraction

- 2k axial or star runs (k = # of independent variables)

- Center runs

The resolution V fraction contributes to the estimation of the linear terms and two-factor interactions. It is variance-optimal for these terms. The axial points contribute to the estimation of the quadratic terms. The center runs provide an internal estimate of error (pure error) and contribute toward the estimation of quadratic terms. Since the phase one experiments required the fractional factorial design and the center runs, to complete the CCD the axial runs were all that was required.

The areas of flexibility in the use of the CCD resides in the selection of α, the axial distance, and the number of center runs. According to Myers and Montgomery [11] and Box and Draper [3], the CCD that is most effective from a variance point of view is to use $\alpha = \sqrt{k}$ and three to five center runs. This design is not necessarily rotatable but is near-rotatable. Therefore, the four center runs completed in the initial augmented fractional factorial design were sufficient for the CCD, and the 10 additional axial runs at $\alpha = \sqrt{5} = 2.236$ were all that were required to complete the CCD.

Once all runs were complete, a mechanism for choosing the values of the indendent variables that would result in the most-optimal mission effectiveness for all responses had to be determined. Because of the complexity and multi-dimensionality of the response surfaces, a simple overlaying of contour plots to graphically choose the point which appeared to be optimal over all responses was not applicable. Therefore, the Derringer and Suich [4] desirability function for optimizing over multiple responses was employed. This method allows for the creation of desirability functions (d_1, d_2, d_3, d_4) for each response where the desirability function can target a specific value, minimize or maximize a response. Since all the responses in this study were measures of mission effectiveness, the desirability functions used were all maximizing. Then a single composite response (D) is developed which is a weighted mean of the desirabilities of the individual responses. The weights in the composite response allow more emphasis on specific individual responses when computing the single composite response. In this study, extra emphasis was placed on two of the responses: the number of unique kills and the number of hits on priority one targets (target formula).

In order to find the optimal conditions using this method, each of the individual desirability functions (d_1, d_2, d_3, d_4) and the composite desirability function (D) must be computed at each design point according to the individual responses. Then a response surface must be built with the computed response D, and appropriate methodology must be applied for finding the conditions that maximize $\hat{D}$ (the model that provides the expected values of D). Since this method will result in many possible combinations of independent variables that will "optimize" the overall mission effectiveness, many of the various combinations must be applied and compared in order to choose the "optimal" settings.

5.   Results and analysis

5.1.   Quantitative results and analysis

Specific numerical results are shown for four scenarios where a cooperative engagement decision algorithm employing the optimal settings resulted in overall improvement over baseline (non-cooperative) performance. Each scenario was defined by three general characteristics:

  1 Warhead lethality/guidance precision

  2 ATR precision

  3 Battlefield characteristics

The specific parameters that were varied in the simulation to define the three general characteristics above were:

  1 Probability of Kill (P_k)

2 False target attack rate (α) and probability of target report (P_{TR})

3 Clutter density (η) and whether the targets were clustered or widely dispersed

The battlefield used for all simulations was approximately 300 square kilometers in size. Two groups of four submunitions each (totaling eight submunitions) where employed in all scenarios. Each of the groups flew a serpentine pattern that covered the entire battlefield in approximately 20 minutes. Each scenario had a total of eight real targets (three high priority and five low priority). Also, two non-targets were employed in the vicinity of the real targets and a battlefield η of 0.05 per square kilometer were randomly placed throughout the battlefield in all scenarios.

Table 6.4 shows the parameters defining scenario 1. This submunition has a relatively non-lethal warhead and is searching for targets clustered in a four square kilometer region of the battlefield.

Table 6.4. Scenario 1 Defining Parameters

Parameter	Value
P_k	0.5
α	0.0053 per square km
P_{TR}	0.95
Target Layout	Cluster

The RSM described in section 4 was performed on this scenario to determine the ideal weighting parameters for the decision rule (equation (2)). When performing the RSM, each simulation run required was reported as a summary of 200 Monte Carlo runs. Each repetition was completed using a different baseline seed for the Monte Carlo simulation. The resulting parameters are shown in Table 6.5.

Table 6.5. Ideal Parameters for Scenario 1 Decision Algorithm

Variable	Weight on	Ideal Value
α_1	Time of Flight	0.77
α_2	Target Priority	0.14
α_3	Range Rate	0.35
α_4	Number of Munitions	0.0
	Maximum Communications Range	9.8 km

The expected performance of the wide area search munitions employing the decision algorithm with the ideal weighting parameters and ideal maximum communications range was then compared to their baseline performance (no cooperation). Table 6.6 shows these results for each of the responses. The overall percent improvement is simply an average of the percent improvements corresponding to each of the four responses.

Table 6.6. Scenario 1 Results

Response	No Cooperation	Ideal Cooperation	Improvement
Unique Kills	2.7	2.81	4.07%
Total Kills	3.06	3.37	10.13%
Total Hits	6.08	6.515	7.15%
Formula	8.04	8.72	8.46%
Overall			7.45%

Table 6.7 shows the parameters defining scenario 2. This submunition has a *lethal* warhead and is searching for targets clustered in a four square kilometer region of the battlefield (same battlefield as scenario 1).

Table 6.7. Scenario 2 Defining Parameters

Parameter	Value
P_h	0.8
α	0.0053 per square km
P_{TR}	0.95
Target Layout	Cluster

The RSM described in section 4 was performed on this scenario to determine the ideal weighting parameters for the decision rule (equation 2) in a similar manner to that for scenario 1. The resulting parameters are shown in Table 6.8.

The same performance measurements as in scenario 1 were analyzed for this scenario. Table 6.9 shows these results for each of the responses.

Table 6.10 shows the parameters defining scenario 3. This submunition has a relatively non-lethal warhead and is searching for targets widely dispersed throughout the entire battlefield.

The same RSM as the previous scenarios was performed on this scenario. The resulting parameters are shown in Table 6.11.

Table 6.12 shows the results for each of the responses.

Table 6.8. Ideal Parameters for Scenario 2 Decision Algorithm

Variable	Weight on	Ideal Value
α_1	Time of Flight	0.30
α_2	Target Priority	0.36
α_3	Range Rate	0.42
α_4	Number of Munitions	0.0
	Maximum Communications Range	20.3 km

Table 6.9. Scenario 2 Results

Response	No Cooperation	Ideal Cooperation	Improvement
Unique Kills	4.13	4.18	1.21%
Total Kills	4.95	5.25	6.06%
Total Hits	6.145	6.53	6.27%
Formula	8.11	8.77	8.11%
Overall			5.42%

Table 6.10. Scenario 3 Defining Parameters

Parameter	Value
P_k	0.5
α	0.0053 per square km
P_{TR}	0.95
Target Layout	Widely Dispersed

Table 6.13 shows the parameters defining scenario 4. This submunition has a *lethal* warhead and is searching for targets that are widely dispersed throughout the entire battlefield (same battlefield as scenario 3).

The resulting parameters after completing the RSM are shown in Table 6.14.

Table 6.15 shows the results for each of the responses.

5.2.   Qualitative results and analysis

As the precision of the ATR is degraded and/or the clutter density increases, this form of cooperative engagement does not offer any advantages and often deteriorates the overall performance of the wide area search munitions. This is because of the hyper-sensitivity to α, the false target attack rate. By degrading the ATR precision and/or increasing

Table 6.11. Ideal Parameters for Scenario 3 Decision Algorithm

Variable	Weight on	Ideal Value
α_1	Time of Flight	0.71
α_2	Target Priority	0.48
α_3	Range Rate	0.1
α_4	Number of Munitions	0.1
	Maximum Communications Range	13.3 km

Table 6.12. Scenario 3 Results

Response	No Cooperation	Ideal Cooperation	Improvement
Unique Kills	2.72	2.70	-0.74%
Total Kills	3.07	3.35	9.12%
Total Hits	6.295	6.52	3.57%
Formula	8.56	9.245	8.00%
Overall			4.99%

Table 6.13. Scenario 4 Defining Parameters

Parameter	Value
P_k	0.8
α	0.0053 per square km
P_{TR}	0.95
Target Layout	Widely Dispersed

the clutter density, α increases. Therefore, what often occurs is that a submunition falsely identifies a clutter or non-target as a real target and then communicates to some of the other munitions the existence of a *real*-target that doesn't actually exist. Then one or more of the other submunitions will decide to cooperatively engage that false target. Now the best event that could occur for that redirected submunition is that it just happens to encounter a real target on its flight path to the false target (the chances of that event occurring being no better than if the submunition would have just stayed on its original search pattern). However, if that does not happen, the submunition is guaranteed to encounter that false target that it thinks is a real target. Upon encountering the false target, the submunition may correctly identify it and not engage it, but as α increases, this is less and less likely. Therefore, cooperative

Table 6.14. Ideal Parameters for Scenario 4 Decision Algorithm

Variable	Weight on	Ideal Value
α_1	Time of Flight	0.31
α_2	Target Priority	0.35
α_3	Range Rate	0.40
α_4	Number of Munitions	0.15
	Maximum Communications Range	19.7 km

Table 6.15. Scenario 4 Results

Response	No Cooperation	Ideal Cooperation	Improvement
Unique Kills	3.93	3.99	1.53%
Total Kills	4.97	5.3	6.64%
Total Hits	6.225	6.555	5.30%
Formula	8.38	9.03	7.76%
Overall			5.31%

engagement alone cannot overcome the hyper-sensitivity in wide area search munition effectiveness to increasing α.

What happens when α remains low, but P_{TR} decreases? This means that given real target encounters, the probability that the ATR is correctly identifying the real targets is decreasing, i.e., there is an increase in submunitions not engaging real targets because they are falsely identifying them as non-targets. In this situation, as long as α remains low, cooperation can still improve overall effectiveness. This is because a submunition may later encounter and correctly identify (and therefore communicate and engage) a real target that another submunition may have previously incorrectly identified as a false target. Then through cooperation, the submunition that originally made the incorrect identification could go back and get a second look at that target and possibly correctly identify and engage it. A scenario such as this will also benefit from redundant area coverage with the initial search patterns at the expense of reduced total area coverage rate.

5.3. Robustness and sensitivity analysis

To test the robustness of the optimal decision parameters determined for each scenario, the optimal decision rule for one scenario was run on a different scenario and then compared to the baseline performance. For example, the optimal decision parameters for scenario 1 (as defined in

Table 6.5) were implemented in the simulation setup to run scenario 2 (as defined by the parameters listed in Table 6.7). This was done for all combinations of the four scenarios described in the quantitative results section (section 5.1). The results proved very little robustness to the selection of the optimal decision parameters. In most cases, the performance with the sub-optimal decision parameters resulted in a zero to two percent overall improvement over baseline performance, but sometimes resulted in deteriorated performance when compared to the baseline.

With these results an attempt was then made to correlate the values of the optimal weighting parameters to the parameters used to define the different scenarios. However, due to the diversity in the optimal weighting parameters across all four scenarios, very little correlation was recognized. The only parameter that displayed some sort of consistency was that associated with the fuel remaining or time of flight–there appears to be some value in waiting until the latter part of the search pattern to choose to cooperatively engage a known target. This, of course, makes sense and allows for the greatest exploration of the entire battlefield.

6.   Conclusions and recommendations

The methods used in this research are not limited to any particular type of wide area search munition and were consciously completed using parameters that describe a very generic wide area search munition. This research, therefore, applies to all wide area search munitions and other cooperative vehicles, and more specific results can be achieved for any specific system by simply modifying the parameters in the effectiveness simulation. Further, the methods developed as part of this research have applications in the more general area of cooperative behavior and control.

The form of cooperative engagement used in this study is most useful in overcoming the limitations on warhead lethality and, to a lesser degree, P_{TR}. However, cooperative engagement alone is not able to compensate for higher false target attack rates. Also, the selection of the optimal weights in the decision algorithm are very sensitive to all battlefield characteristics.

To improve these results, research on cooperative search and cooperative discrimination must be included with the cooperative engagement algorithm to better achieve the full synergistic value of cooperative wide area search munitions.

References

[1] Arkin, R.C. "Cooperation Without Communication: Multiagent Schema-Based Robot Navigation," *Journal of Robotic Systems*, *9*(3):351–364 (1992).

[2] Asama. "Distributed Autonomous Robotic System Configured With Multiple Agents and Its Cooperative Behaviors," *Journal of Robotics and Mechatronics*, *4*(3):199–204 (1992).

[3] Box, G.E.P. and N.R. Draper. *Empirical Model-Building and Response Surfaces*. Applied Probability and Statistics, New York: John Wiley and Sons, 1987.

[4] Derringer, G. and R. Suich. "Simultaneous Optimization of Several Response Variables," *Journal of Quality Technology*, *12*:214–219 (1980).

[5] Dorigo, M. "Ant Colony Optimization." http://iridia.ulb.ac.be/ mdorigo/ACO/ACO.html, October 2000.

[6] Frelinger, David, et al. *Proliferated Autonomous Weapons; An Example of Cooperative Behavior*. Technical Report, RAND, 1998.

[7] Jacques, David R. and Robert Leblanc. "Effectiveness Analysis for Wide Area Search Munitions," *American Institute of Aeronautics and Astronautics, Missile Sciences Conference*, *1*(1):1 (1998 1998). Monterey, CA.

[8] Kube, Ronald C. and Hong Zhang. "Collective Robots: From Social Insects to Robots," *Adaptive Behavior*, *2*(2):189–218

[9] Kwok, Kwan S. and Brian J. Driessen. "Cooperative Target Convergence Using Multiple Agents," *SPIE*, *3209*:67–75 (1997).

[10] Lockheed Martin Vought Systems. "Powered Submunition (PSUB) Effectiveness Model."

[11] Myers, Raymond H. and Douglas C. Montgomery. *Response Surface Methodology: Process and Product Optimization Using Designed Experiment*. Wiley Series in Probability and Statistics, New York: Wiley-Interscience, 1995.

[12] Reynolds, C.W. "Flocks, Herds, and Schools," *Computer Graphics*, *21*(4):25–34 (July 1987).

Chapter 7

ON A GENERAL FRAMEWORK TO STUDY COOPERATIVE SYSTEMS

Victor Korotkich
Faculty of Informatics and Communication
Central Queensland University
Mackay, Queensland 4740, Australia
v.korotkich@cqu.edu.au

Abstract   A collection of many systems that cooperatively solve an optimization problem is considered. The consideration aims to determine criteria such that the systems as a whole show their best performance for the problem. A general framework based on a concept of structural complexity is proposed to determine the criteria. The main merit of this framework is that it allows to set up computational experiments revealing the criteria. In particular, the experiments give evidence to suggest that criteria of best performance are realized when the structural complexity of cooperative systems equals the structural complexity of the optimization problem. The results of the paper could give a new perspective in the developing of optimization methods based on cooperative systems.

Keywords: cooperative systems, best performance, optimization, complexity, computational phase transitions, traveling salesman problem

1.   Introduction

A collection of many systems that cooperatively solve an optimization problem is considered. The consideration aims to determine criteria such that the systems as a whole show their best performance for the problem. The dynamics of each system is rules-based and rules providing the criteria are sought. In a broad sense the consideration addresses a fundamental problem: given a number of systems how to form by using them a new system with properties of interest.

R. Murphey and P.M. Pardalos (eds.), Cooperative Control and Optimization, 121–142.
© 2002 *Kluwer Academic Publishers.*

Many research fields are intensively studied now to deal with complex cooperative phenomena (for example [1] - [5]). However, despite impressive results in the fields, a general framework that could identify the criteria is missing.

In the paper, a general framework based on a concept of structural complexity is proposed to determine the criteria. The main merit of this framework is that it allows to set up computational experiments revealing the criteria. In particular, the experiments give evidence to suggest that criteria of best performance are realized when the structural complexity of cooperative systems equals the structural complexity of the optimization problem.

Natural systems in space-time can be described in terms of the energy functions. This approach is very much accustomed for many physical systems whose energy functions are available because interactions are known properly. It is a different situation in our case. The approach has to deal with cooperative systems whose interactions cannot be so generally defined and known as in physics. Another substantial difficulty is that the space-time approach has no means to capture a specific optimization problem and translate this information into a form expressing the criteria for the problem. In other words, the language of space-time, i.e., energy and related notions, is not adequate to characterize optimization problems. It is well-known that concept of complexity is more appropriate for their description.

In this paper, instead of developing a general mechanism of energy functions construction for cooperative systems and make it then coherent with language of complexity, a different approach is tried. We propose a general framework within which both cooperative systems and complexity find universal description in terms of hierarchical formations of integer relations. The general framework uses laws of these hierarchical formations to analyse and control cooperative systems as well as to measure complexity of natural systems.

In section 1, a concept of structural complexity and its relevance to the subject of the paper are briefly given. Structural complexity is defined to measure complexity of natural systems in terms of hierarchical formations of integer relations [6].

By using the concept of structural complexity an optimal algorithm for $N = 2$ two cooperative systems is proposed in [7]. The optimal algorithm can be formulated as a simple strategy: "win - stay, lose - consult PTM generator" of the celebrated Prouhet-Thue-Morse (PTM) sequence [8]-[10]. In section 2, to explore the optimal algorithm for $N > 2$ cooperative system its parameter extension is presented.

In section 3, the parameter extension is used to investigate the relevance of the optimal algorithm to the criteria. Extensive computational experiments are used in the investigation as its difficulty gives small chances for theoretical means in the first place. For N cooperative systems the parameter extension is formed into an approximate method for solving the travelling salesman problem (TSP). The class of benchmark TSP problems [11] is used in the computational experiments.

The experiments exhibit computational phase transitions. Namely, the performance of cooperative systems for values of parameters corresponding to the optimal algorithm improves significantly with effects out of proportion to cause. It turns out that the optimal algorithm is the critical point in the parameter extension. In this point cooperative systems as a whole show their best performance for a TSP problem thus demonstrating that the optimal algorithm is relevant to the criteria.

In section 4, the above results motivate us to develop and use further the concept of structural complexity as the key in determining the criteria.

Firstly, a concept of structural complexity for N cooperative systems is introduced. The concept specifies a space, called the complexity space, whose elements are so called complexity matrices. The coordinate system in the complexity space is based on a partial order that captures our intuitive understanding of the order of successive positions resulting in a monotone increase of structural complexity. As a collection of cooperative systems moves along such a trajectory its structural complexity increases monotonically. The key point of our interest is to know how such a motion with monotonic increase in structural complexity affects performance of cooperative systems to solve an optimization problem.

Secondly, a parameter controlling monotonic increase of structural complexity is proposed. The parameter is used in computational experiments to investigate how average performance of cooperative systems is connected with their structural complexity as it grows monotonically. The experiments show a remarkable fact that for each optimization problem tested there is an unique point on a trajectory in the complexity space such that the performance of cooperative systems monotonically increases till this point and then monotonically decreases after it.

Thirdly, such points make it possible to associate an optimization problem with corresponding complexity matrices and use them to define structural complexity of the optimization problem. Finally, this allows us to suggest that the criteria of best performance are realized when the structural complexity of cooperative systems equals the structural complexity of the optimization problem.

In conclusion, we summarize results of the paper and briefly discuss possible relevant questions in the future study of cooperative systems. In particular, the results of the paper could give a new perspective in the developing of optimization methods based on cooperative systems. The optimization methods according to the criteria should be devised to have parameters controlling their structural complexity handy. The use of the methods for an optimization problem could be interpreted as a process in which the parameters must be tuned in an optimal way to get the criteria.

2.   Structural complexity

In this section, a concept of structural complexity and its relevance to the subject of the paper are briefly given.

Let I be an integer alphabet and

$$I_n = \{s = s_1...s_n, s_i \in I, i = 1, ..., n\}$$

be the set of all sequences of length $n \geq 2$ with symbols in I. If $I = \{-1, +1\}$ then I_n is the set of all binary sequences of length n, denoted B_n. Let $s(i) = s_1...s_i$ and $|s(i)| = i$ denote the length of $s(i) \in I_i, i = 1, ..., n$.

If for a pair of different sequences $s = s_1...s_n \in I_n$, $s' = s'_1...s'_n \in I_n$, $s \neq s'$ there is an integer $k \geq 2$ such that their first $(k-1)$, so called structural numbers are equal

$$\vartheta_1(s) = \vartheta_1(s'), ..., \vartheta_{k-1}(s) = \vartheta_{k-1}(s'),$$

whereas the kth structural numbers are not equal $\vartheta_k(s) \neq \vartheta_k(s')$ then

$$C(s, s', n) = k,$$

where

$$\vartheta_k(s) = \sum_{i=0}^{k-1} \alpha_{kmi}((m + n)^i s_1 + ... + (m + 1)^i s_n),$$

$\alpha_{kmi}, i = 0, ..., k - 1$ are rational coefficients and m is an integer (for more details see [6]). By definition $\vartheta_0(s) = 0$ and $C(s, s, n) = 0$, $s \in I_n$.

It is shown that $C(s, s', n) = k \leq n$ [6] leading to a system of $(k-1)$ linear equations in integers $(s_i - s'_i)$, $i = 1, ..., n$

$$(m + n)^0(s_1 - s'_1) + ... + (m + 1)^0(s_n - s'_n) = 0$$

$$\cdot \qquad \cdot \qquad \cdot \qquad \cdot$$

$$(m + n)^{k-2}(s_1 - s'_1) + ... + (m + 1)^{k-2}(s_n - s'_n) = 0 \qquad (1)$$

and inequality

$$(m+n)^{k-1}(s_1 - s_1') + ... + (m+1)^{k-1}(s_n - s_n') \neq 0. \qquad (2)$$

The system (1) can be viewed as a system of integer relations. The key idea concerning the system (1) and inequality (2) is that they represent hierarchical formations, i.e., hierarchical formations of integer relations $WR(s, s', n, m, I_n)$ and two-dimensional geometrical patterns $WP(s, s', n, m, I_n)$, called integer patterns(for more details see [6] and Figure 7.1 as an illustration).

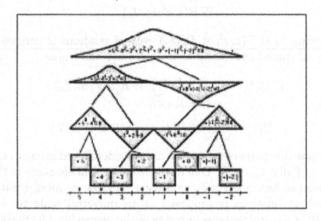

Figure 7.1. The figure shows a hierarchical formation of integer relations revealing laws and the process of the formation. The element of the highest level, i.e., integer relation $+5^2 - 4^2 - 3^2 + 2^2 - 1^2 + 0^2 + (-1)^2 + 2^2 = 0$, if associated with a system, through this figure would give a lot of information about the structure of the system and its complexity. The structural complexity of this hierarchical formation is 4, i.e., the highest level of it.

By definition

$$WR(s, s, n, m, I_n) = \emptyset, \quad WP(s, s, n, m, I_n) = \emptyset, \quad s \in I_n.$$

It is shown that the hierarchical formation of integer relations $WR(s, s', n, m, I_n)$ is isomorphic to the hierarchical formation of integer patterns $WP(s, s', n, m, I_n)$

$$\left\{ \begin{array}{c} \textit{System of Linear Equations (1)} \\ \textit{and Inequality (2)} \end{array} \right\}$$

$$\Downarrow$$

$$\left\{ \begin{array}{c} \textit{Hierarchical Formation} \\ \textit{of Integer Relations} \end{array} \right\} \underset{\textit{isomorphism}}{\Longleftrightarrow} \left\{ \begin{array}{c} \textit{Hierarchical Formation} \\ \textit{of Integer Patterns} \end{array} \right\}$$

All hierarchical formations of integer relations associated with sequences s, s' and their initial segments $s(i), s'(i)$ of length $i = 1, ..., n-1$ are incorporated into one hierarchical formation

$$WR(s, s', m, I_n) = \bigcup_{i=1}^{n} WR(s(i), s'(i), i, m, I_i).$$

The isomorphism specifies a hierarchical formation of integer patterns

$$WP(s, s', m, I_n)$$

corresponding to $WR(s, s', m, I_n)$. A web of relations is defined to incorporate all these hierarchical formations into one whole

$$WR(I_n) = \bigcup_{m \in \mathbf{Z}} \bigcup_{s \in I_n} \bigcup_{s' \in I_n} WR(s, s', m, I_n),$$

$$WR(I) = \lim_{n \to \infty} WR(I_n), \quad WR = WR(\mathbf{Z}),$$

where all possible pairs of sequences are considered and integer alphabet I is the set of all integers $\mathbf{Z}$. Due to the isomorphism elements of the web of relations can be equally seen as integer relations or integer patterns.

The web of relations is proposed as an universal scale to measure complexity of natural systems in terms of the hierarchical formations [6]. The following concept of complexity is developed in this direction.

Structural Complexity C(s,s'). *Structural complexity $C(s, s')$ of a sequence $s \in I_n$ with respect to a sequence $s' \in I_n$, $s \neq s'$ is defined by*

$$C(s, s') = \max_{i=1,...,n} C(s(i), s'(i), i)$$

and $C(s, s) = 0$, $s \in I_n$.

We briefly explain why structural complexity becomes our basic concept. The subject of the paper is connected with a fundamental problem: given a number of systems how to form by using them a new system with properties of interest. The general framework uses laws of the hierarchical formations to analyse and control cooperative systems. Structural complexity plays the leading role because it is a very important means to characterize these hierarchical formations. The structural complexity $C(s, s')$ of hierarchical formations $WR(s, s, m, I_n)$ and $WP(s, s', m, I_n)$ equals the maximum level the hierarchical formations progress to. They form at each next level more complex elements as compositions of elements of the lower level.

The general framework is a final theory [12], as it is based on irreducible concepts only, i.e., integers. The framework turned out to be true for cooperative systems, there would be no need for a deeper theory.

3. Parameter extension of the optimal algorithm

By using a concept of structural complexity an optimal algorithm for $N = 2$ two cooperative systems is proposed in [7]. The optimal algorithm can be formulated as a simple strategy: "win - stay, lose - consult PTM generator" of the celebrated Prouhet-Thue-Morse (PTM) sequence. To explore the optimal algorithm for $N > 2$ cooperative system its parameter extension is presented in this section.

The optimal algorithm is found as the solution to a problem of binary sequence prediction with structural complexity as the criterion. In the problem a bit s_k must be produced at each kth $k = 1, ..., n$ step $n \geq 1$ as the prediction of the kth bit s'_k of a binary sequence s' before this bit s'_k becomes actually known. After such n steps a sequence $s = s_1...s_n$ is produced and compared with the sequence $s' = s'_1...s'_n$.

What matters in the comparison is the structural complexity $C(s, s')$ of the sequence s with respect to the sequence s'. The prediction problem is to maximize on average the structural complexity $C(s, s')$ given that the binary sequences of length n are all equally possible as input s' for the prediction.

Let $G(\eta^+)$ be a generator of the PTM sequence starting with $+1$

$$\eta^+ = +1 - 1 - 1 + 1 - 1 + 1 + 1 - 1 - 1 + 1 + 1 - 1 + 1 - 1 - 1 + 1 \ldots$$

consequently producing one bit of the sequence per iteration. With the help of the generator $G(\eta^+)$ the optimal algorithm

$$\Lambda_2^* = \{\chi_k^*, \ k = 1, ..., n\},$$

where χ_k^* is the prediction rule at the kth step $k = 1, ..., n$, admits the following

Chaotic description of the optimal algorithm

Step 1. Set the number of steps $k = 1$ and number of incorrect steps $l = 1$.

Step 2. Generate the lth bit η_l^+ of the PTM sequence η^+ by the generator $G(\eta^+)$ and set $\chi_k^* = \eta_l$. Set $s_k = \chi_k^*$.

Step 3. If the step is correct, i.e., $s_k = s'_k$, then go to Step 4. Otherwise, if the step is incorrect, i.e., $s_k = -s'_k$, go to Step 5.

Step 4. If $k = n$, then go to Step 6. Otherwise set $\chi_{k+1}^* = \chi_k^*$ and $s_{k+1} = \chi_{k+1}^*$. Increment the number of steps $k = k + 1$ and go to Step 3.

Step 5. If $k = n$, then go to Step 6. Otherwise increment the number of steps $k = k + 1$ and the number of incorrect steps $l = l + 1$ and go to Step 2.

Step 6. Stop.

The algorithm preserves the optimality if it uses a generator of the PTM sequence starting with -1

$$\eta^- = -1 + 1 + 1 - 1 + 1 - 1 - 1 + 1 + 1 - 1 - 1 + 1 - 1 + 1 + 1 - 1 \ldots$$

instead of the generator $G(\eta^+)$.

Algorithm A_2^* is optimal in terms of structural complexity for $N = 2$ two cooperative systems only. To explore the optimal algorithm for $N > 2$ cooperative system a parameter extension of the algorithm is presented.

The parameter extension, denoted $\mathbf{A}_2^*(\bar{p})$, is defined by incorporating parameters $\bar{p} = (p_1, p_2, p_3)$ into the chaotic description of the algorithm and can be viewed as a set of algorithms with each element, i.e., algorithm,

$$A_2^*(\bar{p}) = \{\chi_k^*(\bar{p}), \ k = 1, ..., n\}.$$

identified by specific values of the parameters $\bar{p} = (p_1, p_2, p_3)$. The parameters have the following meaning:

1. $p_1 \geq 1$ - a parameter that specifies how many steps $A_2^*(\bar{p})$ can proceed sequentially with the same decision if it is correct all the time. Suppose, due to $G(\eta^+)$, a new decision is made by $A_2^*(\bar{p})$ and it is correct for p_1 successive steps. Then $A_2^*(\bar{p})$ does not follow the decision but instead refers to the PTM generator $G(\eta^+)$ for the next bit of the sequence η^+ to make a decision;

2. $p_2 \geq 1$ - a parameter that specifies how many steps $A_2^*(\bar{p})$ can proceed sequentially with the same decision if it is incorrect all the time. Suppose, due to $G(\eta^+)$, a new decision is made by $A_2^*(\bar{p})$ and it is incorrect for p_2 successive steps. Then $A_2^*(\bar{p})$ does not follow the decision but instead refers to the PTM generator $G(\eta^+)$ for the next bit of the sequence η^+ to make a decision;

3. $p_3 \geq 1$ - a parameter that specifies a bit of the PTM sequence η^+ with which $A_2^*(\bar{p})$ starts using the sequence. For the optimal algorithm A_2^* we have $p_3 = 1$.

The idea of using parameters p_1, p_2 is that they can control the structural complexity of sequences producing by the parameter extension $\mathbf{A}_2^*(\bar{p})$. Parameter p_3 is used to diversify the behaviour of $\mathbf{A}_2^*(\bar{p})$.

Description of algorithm in the parameter extension

Step 1. Set the number of steps $k = 1$, number of incorrect steps $l = 1$, number of current correct steps $k' = 0$ and the number of current incorrect steps $l' = 0$.

Step 2. Generate at random the $(p_3 + l - 1)$th bit η_{p_3+l-1} of the PTM sequence η^+ by the generator $G(\eta^+)$ and for the value of the rule $\chi_k^*(\bar{p})$ set $\chi_k^*(\bar{p}) = \eta_{p_3+l-1}$. Set $s_k = \chi_k^*(\bar{p})$.

Step 3. If the step is correct, i.e., $s_k = s_k'$, then increment the number of current correct steps $k' = k' + 1$ and go to Step 4. Otherwise, if the step is incorrect, i.e., $s_k = -s_k'$, go to Step 5.

Step 4. If $k = n$, then go to Step 6. Otherwise if $k' < p_1$ then set $\chi_{k+1}^*(\bar{p}) = \chi_k^*(\bar{p})$ and $s_{k+1} = \chi_{k+1}^*(\bar{p})$. Increment the number of steps $k = k + 1$ and go to Step 3. If $k' = p_1$ then increment the number of steps $k = k + 1$, number of incorrect steps $l = l + 1$, set $k' = 0$ and go to Step 2.

Step 5. If $k = n$, then go to Step 6. Otherwise if $l' < p_2$, then set $\chi_{k+1}^*(\bar{p}) = \chi_k^*(\bar{p})$ and $s_{k+1} = \chi_{k+1}^*(\bar{p})$. Increment the number of steps $k = k + 1$, set $l' = l' + 1$ and go to Step 3. If $l' = p_2$ then increment the number of steps $k = k + 1$, number of incorrect steps $l = l + 1$, set $l' = 0$ and go to Step 2.

Step 6. Stop.

Optimal algorithm A_2^* in the set $\mathbf{A}_2^*(\bar{p})$ is specified by the conditions

$$p_1 = n, \quad p_2 = 1, \quad p_3 = 1.$$

4. Critical point in the parameter extension: the optimal algorithm

The parameter extension in this section is used to investigate the relevance of the optimal algorithm to the criteria. Extensive computational experiments are used in the investigation as its difficulty gives small chances for theoretical means in the first place. For N cooperative systems the parameter extension is formed into an approximate method for solving the travelling salesman problem (TSP). The class of benchmark TSP problems [11] is used in the computational experiments.

The experiments exhibit computational phase transitions. Namely, the performance of cooperative systems for values of parameters corresponding to the optimal algorithm improves significantly with effects out of proportion to cause. It turns out that the optimal algorithm is the critical point in the parameter extension. In this point cooperative systems as a whole show their best performance for a TSP problem thus demonstrating that the optimal algorithm is relevant to the criteria.

The approximate method admits the following description. In the description it is convenient to view a collection of N cooperative systems as a group of N people who cooperatively solve the TSP problem, i.e., to find the shortest closed path between $n+2$ cities going through each city exactly once. The group of people in solving a TSP problem proceeds in the following manner. They all start in the same city and each of them first chooses the next city at random. Then at each city the person has two choices, he or she may pick up a next city purely at random or travel to the next city which is currently closest to him or her.

In other words, at each city the person makes a decision to use a random algorithm or the greedy one in order to get to the next city. A decision of the person to use a random algorithm or the greedy one can be denoted by $+1$ or -1 respectively. This associates with the person a binary sequence of length n as he or she step by step visits all cities.

People of the group cooperate with each other while solving the TSP problem. After each step distances travelled by all of them so far become known to everyone and this affects their decisions. Namely, the person, before making a next decision, evaluates if his or her last decision is correct or incorrect by using a parameter p_4. If according to the information the person finds out that the distance he or she travelled so far deviates from the shortest one to date by not more than $p_4 \geq 0\%$ percent then his or her last decision is correct and incorrect otherwise. Therefore, decisions of a person somehow depend on the decisions made by other people of the group and in its turn their decisions depend on the decisions made by the person.

In the method a person of the group makes decisions by using the parameter extension $\mathbf{A}_2^*(\bar{p})$, $\bar{p} = (p_1, p_2, p_3, p_4)$. The performance of the group with the ith $i = 1, ..., N$ person using an algorithm $A_2^*(\bar{p}_i)$ (values of the parameters are fixed) is characterised by

$$D(\bar{\mathbf{p}}) = \min_{i=1,...,N} D_i(\bar{p}_i),$$

where $\bar{\mathbf{p}} = (\bar{p}_1, ..., \bar{p}_N)$, $\bar{p}_i = (p_1, p_2, p_3, p_4)$, $i = 1, ..., N$ and $D_i(\bar{p}_i)$ is the total distance traveled by the ith person after visiting all cities. The distance $D(\bar{\mathbf{p}})$ emerges from the cooperative search of the whole group and is based on the totality of decisions made by the people. It characterises the cooperative ability of the group to find the shortest path D^* in terms of how close $D(\mathbf{p})$ approaches D^*.

Extensive computational experiments to understand the character of $D(\mathbf{p})$ have been made by using the class of benchmark TSP problems [11]. In the experiments $N = 10, 20, ..., 200, 250, ..., 1000, 2000, ..., 5000$, the parameters have been within ranges $p_1 = 1, 5, 10, ..., n$, $p_2 = 1, 5, 10, ..., n$, $p_3 = 1, 2, ..., N$ and $p_4 = 0\%, 1\%, ..., 20\%$.

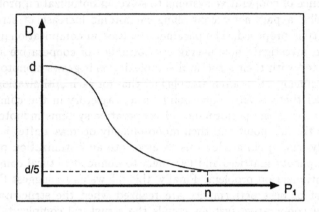

Figure 7.2. The figure schematically summarizes the typical situation observed in the computational experiments. The minimal distance drops very significantly as parameters approach values of the optimal algorithm.

The experiments reveal that parameter p_1 plays a very important role in improving the performance of the method. In particular, $D(\mathbf{p})$, when averaged over all possible values of the other parameters, for all the TSP problems suddenly drops to less then $\frac{1}{5}$ of its value as parameter p_1 approaches n. Figure 7.2 schematically summarizes the typical situation with the parameter p_1. This can be interpreted as a continuous (second order) computational phase transition [13]. The other parameters do not have such a noticeable effect, but as parameter p_2 tends to 1 a much better performance is definitely observed.

5. Structural complexity of cooperative systems versus optimization problems

In this section, the above results motivate us to develop and use further the concept of structural complexity as the key in determining the criteria.

Firstly, a concept of structural complexity for N cooperative systems is introduced. The concept specifies a space, called the complexity space, whose elements are so called complexity matrices. The coordinate system in the complexity space is based on a partial order that captures our intuitive understanding of the order of successive positions resulting in a monotone increase of structural complexity. As a collection of cooperative systems moves along such a trajectory its structural complexity increases monotonically. The key point of our interest is to know how such a motion with monotonic increase in structural complexity affects performance of cooperative systems to solve an optimization problem.

Secondly, a parameter controlling monotonic increase of structural complexity is proposed. The parameter is used in computational experiments to investigate how average performance of cooperative systems is connected with their structural complexity as it grows monotonically. The experiments show a remarkable fact that for each optimization problem tested there is an unique point on a trajectory in the complexity space such that the performance of cooperative systems monotonically increases till this point and then monotonically decreases after it.

Thirdly, such points allows us to associate an optimization problem with complexity matrices and use them to define structural complexity of the optimization problem. Finally, this allows us to suggest that the criteria of of best performance are realized when the structural complexity of cooperative systems equals the structural complexity of the optimization problem.

5.1. The complexity space

Let $\mathbf{S}(n, N)$ be the set of all equivalence classes of all the $N \times n$ binary matrices with each class containing different matrices as permutations of each other. A collection of N cooperative systems, viewed in terms of the previous section, can be represented in space-time by a $N \times n$ decision matrix

$$S = \{s_{ij}\}, \quad i = 1, ..., N, \ j = 1, ..., n$$

where a binary sequence $\bar{s}_i = s_{i1}...s_{in} \in B_n$ specifies the ith $i = 1, ..., N$ system and s_{ij} is a decision made by the system at the jth step. Let $S \in \mathbf{S}(n, N)$ denote that a decision matrix S belongs to an equivalence class of $\mathbf{S}(n, N)$. All decision matrices of an equivalence class in $\mathbf{S}(n, N)$ are viewed the same.

Natural systems in space-time can be described in terms of the energy functions. For example, one could try to view decision matrices in terms of the Ising model. The space-time approach is very much accustomed for many physical systems whose energy functions are available because interactions are known properly. It is a different situation in our case. The approach has to deal with cooperative systems whose interactions cannot be so generally defined and known as in physics. Another substantial difficulty is that the space-time approach has no means to capture a specific optimization problem and translate this information into a form expressing the criteria for the problem. In other words, the language of space-time, i.e., energy and related notions, is not adequate to characterize optimization problems. It is well known that concept of complexity is more appropriate for their description.

In this paper, instead of developing a general mechanism of energy functions construction for cooperative systems and make it then coherent with language of complexity, a different approach is tried. We propose a general framework within which both cooperative systems and complexity find universal description in terms of hierarchical formations of integer relations. The general framework uses laws of these hierarchical formations to analyse and control cooperative systems as well as to measure complexity of natural systems.

Moreover, a main merit of the hierarchical formation is that it is nothing but *a process of the formation of its elements*. Indeed, what we can observe in a hierarchical formation (see Figure 7.1) is that on each its level elements satisfying certain conditions combine to form more complex elements belonging to the next level. As the result more complex elements than their constituents emerge. The formed elements are based on the elements of the previous level but are more then their simple sum. This character of the hierarchical formation is very close to our mental picture of the problem: given a number of systems to form by using them a new system with properties of interest. This allows our intuition be involved and helpful for the problem.

The collection of N cooperative systems in the web of relations can be represented by a set of hierarchical formations

$$WR(S) = \{WR(\bar{s}_i, \bar{s}_j, 0, B_n),\ \bar{s}_i, \bar{s}_j \in S,\ i, j = 1, ..., N\} \subset WR,$$

where a hierarchical formation $WR(\bar{s}_i, \bar{s}_j, 0, B_n)$ is associated with the ith and jth system $i, j = 1, ..., N$.

These two representations are connected as sequences of the matrix S determine elements of the set $WR(S)$

$$\left\{ \begin{array}{c} Space - time \\ Matrix\ S \end{array} \right\} \Longrightarrow \left\{ \begin{array}{c} Web\ of\ Relations \\ Collection\ of\ Hierarchical\ Formations\ WR(S) \end{array} \right\}$$

A hierarchical formation $WR(\bar{s}_i, \bar{s}_j, 0, B_n)$, $\bar{s}_i \neq \bar{s}_j$ in terms of integer patterns can be viewed as a spatial object giving a static picture of the formation of its elements (see Figure 7.1 for illustration). This object requires many characteristics for its complete specification. A very important characteristic of $WR(\bar{s}_i, \bar{s}_j, 0, B_n)$ is the number of its levels, which, by definition, equals the structural complexity $C(\bar{s}_i, \bar{s}_j)$ of the sequence $\bar{s}_i$ with respect to the sequence $\bar{s}_j$. By using structural complexity $C(\bar{s}_i, \bar{s}_j)$ hierarchical formation $WR(\bar{s}_i, \bar{s}_j, 0, B_n)$ is described in terms of the maximal level or complexity of elements it attains to form.

We propose to describe a set of hierarchical formations $WR(S)$ in terms of a $N \times N$ matrix, called a complexity matrix and defined

$$C(S) = \{C(\bar{s}_i, \bar{s}_j)\}, \ i, j = 1, ..., N, \tag{3}$$

by considering each of its hierarchical formations $WR(\bar{s}_i, \bar{s}_j, 0, B_n) \in WR(S)$ in terms of the structural complexity $C(\bar{s}_i, \bar{s}_j)$. Structural complexity of cooperative systems represented by a decision matrix S is associated with a complexity matrix $C(S)$.

Let $\mathbf{C}(n, N)$ be the set of all $N \times N$ complexity matrices corresponding to $\mathbf{S}(n, N)$ due to (3). Namely, elements of $\mathbf{C}(n, N)$ are equivalence classes with each class containing complexity matrices whose decision matrices belong to the same equivalence class of $\mathbf{S}(n, N)$. Let $C(S) \in \mathbf{C}(n, N)$ denote that a complexity matrix $C(S)$ belongs to an equivalence class of $\mathbf{C}(n, N)$. All complexity matrices of an equivalence class in $\mathbf{C}(n, N)$ are viewed the same.

The principal feature of the set $\mathbf{C}(n, N)$ comes up with introduction of a partial order that captures our intuitive understanding of the order of successive positions resulting in a monotone increase of structural complexity. This partial order converts the set $\mathbf{C}(n, N)$ into a partially ordered set, called the complexity space, and defines the coordinate system allowing to measure structural complexity of cooperative systems.

The partial order is based on the natural assumption that one set of hierarchical formations is more complex than another set of hierarchical formations if structural complexity of elements formed by the first set are greater than structural complexity of elements formed by the second one.

Formally, let $S_1, S_2 \in \mathbf{S}(n, N)$, $S_1 \neq S_2$ and $WR(S_1), WR(S_2) \subset WR$ be two sets of hierarchical formations. The set $WR(S_2)$ is defined to be more complex than the set $WR(S_1)$ if there is a one-to-one correspondence between the sets satisfying:

- the structural complexity of any hierarchical formation in $WR(S_1)$ is not greater than the structural complexity of a corresponding hierarchical formation in $WR(S_2)$;

- there exist at least one hierarchical formation in $WR(S_2)$ such that its structural complexity is greater than the structural complexity of a corresponding hierarchical formation in $WR(S_1)$.

In words, a set of hierarchical formations $WR(S_2)$ is defined more complex than a set of hierarchical formations $WR(S_1)$ if it is like the set

$WR(S_1)$ except at least one of its hierarchical formations produces more complex elements.

The above conditions in terms of complexity matrices appear as follows. Let $C_1, C_2 \in \mathbf{C}(n, N)$, $C_1 \neq C_2$ be two complexity matrices. Matrix C_2 is defined to be more complex than matrix C_2, denoted by $C_1 \prec C_2$ if for all i, j

$$C'_{ij} \leq C''_{ij}$$

and there exists at least one pair of their elements such that

$$C'_{ij} < C''_{ij}.$$

In the complexity space matrices $C_1, C_2, ..., C_k$ linearly ordered by the relation $\prec$, i.e., giving rise to a trajectory

$$C_1 \prec C_2 \prec ... \prec C_{k-1} \prec C_k,$$

are of special interest as each next element in the trajectory is more complex than the current one. As a collection of cooperative systems moves along such a trajectory its structural complexity increases monotonically. The key point of our interest is to know how such a motion with monotonic increase in structural complexity affects performance of cooperative systems to solve an optimization problem.

5.2. Cooperative systems criteria of best performance

Extensive computational experiments have been done to investigate how average performance of cooperative systems is connected with their structural complexity as it grows monotonically. The same class of benchmark TSP problems have been used. As in the case of the parameter extension there are small chances that theoretical means could be helpful on this matter in the first place.

The experiments show a remarkable fact that for each optimization problem tested there is an unique point on a trajectory in the complexity space such that the performance of cooperative systems monotonically increases till this point and then monotonically decreases after it. Such points make it possible to associate an optimization problem with corresponding complexity matrices and use them to define structural complexity of the optimization problem. Finally, this allows us to suggest that the criteria of best performance are realized when the structural complexity of cooperative systems equals the structural complexity of the optimization problem.

To realize the computational experiments we identify a parameter that controls the structural complexity of cooperative systems in a certain way. In particular, due to the optimal algorithm this parameter makes it possible to monotonically increase and decrease it (for more details see [6]).

Let D_{ij} be the distance traveled by the ith $i = 1, ..., N$ person after j steps $j = 1, ..., n$ and

$$D_j^- = \min_{i=1,...,N} D_{ij}, \quad D_j^+ = \max_{i=1,...,N} D_{ij}.$$

All distances D_{ij} traveled by people of the group after j steps lie on the interval $[D_j^-, D_j^+]$. We define a parameter v that specifies a point

$$D_j(v) = D_j^+ - v(D_j^+ - D_j^-), \quad 0 \le v \le 1 + \varepsilon, \quad \varepsilon > 0$$

dividing the interval into two parts. The parameter v determines what people of the group view as a correct decision. If a distance traveled by a person after j steps $j = 1, ..., n$ belongs to the interval $[D_j^-, D_j(v)]$, called successful, then the person's last decision is correct. If the distance traveled by the person belongs to the interval $(D_j(v), D_j^+]$, called unsuccessful, then the person's last decision is incorrect.

The meaning of the parameter v when the optimal algorithm is used can be illustrated by two limiting examples.

1. When $v = 0$, then $D_j(v) = D_j^+$. This means that the successful interval $[D_j^-, D_j(v)]$ coincides with the whole interval $[D_j^-, D_j^+]$ and people of the group at each step always make correct decisions irrespective of distances traveled by them and relations between the distances. Therefore, according to the optimal algorithm a person starting with a decision never changes it all the way and as a result a corresponding binary sequence consists completely of all $+1$s or all -1s depending on initial decision. For example, when $N = 4$, $n = 4$ and all people of the group start with a random algorithm the decision matrix can be written

$$S = \begin{pmatrix} +1 & +1 & +1 & +1 \\ +1 & +1 & +1 & +1 \\ +1 & +1 & +1 & +1 \\ +1 & +1 & +1 & +1 \end{pmatrix} \tag{4}$$

The order in the decision matrix (4) is evident and referring to its interpretation in terms of the Ising model it could be said that it corresponds to the state when all spins are up. The complexity matrix of (4) can be represented

$$C(S) = \begin{pmatrix} 0 & 0 & 0 & 0 \\ 0 & 0 & 0 & 0 \\ 0 & 0 & 0 & 0 \\ 0 & 0 & 0 & 0 \end{pmatrix}$$

2. In the opposite limit when $v = 1 + \varepsilon$, then $D_j(v) < D_j^-$. This means that the unsuccessful interval $(D_j(v), D_j^+]$ covers the whole interval $[D_j^-, D_j^+]$ and people of the group at each step always make incorrect decision irrespective of distances traveled by them and relations between the distances. Therefore, a person following the optimal algorithm at each step always asks PTM generators what decision to make. As these decisions are completely determined by the PTM sequences, a corresponding binary sequence is the PTM sequence starting with $+1$ or -1. For example, when $N = 4$, $n = 4$ and half of the group starts with a random algorithm and the other half starts with the greedy one the decision matrix can be written

$$S = \begin{pmatrix} +1 & -1 & -1 & +1 \\ -1 & +1 & +1 & -1 \\ +1 & -1 & -1 & +1 \\ -1 & +1 & +1 & -1 \end{pmatrix} \tag{5}$$

The complexity matrix of (5) can be represented

$$C(S) = \begin{pmatrix} 0 & 3 & 0 & 3 \\ 3 & 0 & 3 & 0 \\ 0 & 3 & 0 & 3 \\ 3 & 0 & 3 & 0 \end{pmatrix}$$

It is worth to note that for all pairs of sequences $s, s' \in B_4$ but the PTM sequences we have $C(s, s') < 3$. The PTM sequences are assumed to have extreme properties in this respect (see [6] for more details).

The decision matrix (5) has interesting properties, for example a fractal property. By associating

$$\begin{pmatrix} +1 & -1 \\ -1 & +1 \end{pmatrix} \to +1 \qquad \begin{pmatrix} -1 & +1 \\ +1 & -1 \end{pmatrix} \to -1$$

for (5) we come up with

$$\begin{pmatrix} +1 & -1 \\ -1 & +1 \end{pmatrix}$$

which is exactly the left-top (2×2) block of it.

Suppose that half of $N = 2N'$ people $N' = 1, 2, \dots$ of the group starts with a random algorithm and the other half with the greedy one. Then the decision matrix of the group for $v = 0$ can be written

$$S_0 = \begin{pmatrix} +1 & +1 & \dots & +1 & +1 \\ -1 & -1 & \dots & -1 & -1 \\ . & . & \dots & . & . \\ +1 & +1 & \dots & +1 & +1 \\ -1 & -1 & \dots & -1 & -1 \end{pmatrix}$$

and decision matrix for $v = 1 + \varepsilon$ can be written

$$S_1 = \begin{pmatrix} +1 & -1 & ... & \pm 1 & \pm 1 \\ -1 & +1 & ... & \mp 1 & \mp 1 \\ . & . & ... & . & . \\ +1 & -1 & ... & \pm 1 & \pm 1 \\ -1 & -1 & ... & \mp 1 & \mp 1 \end{pmatrix}$$

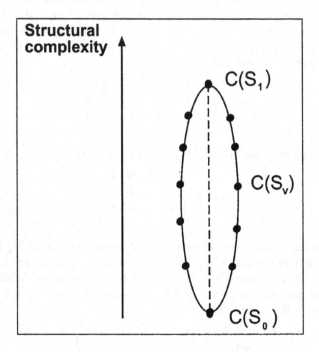

Figure 7.3. Different trajectories connecting matrices $C(S_0), C(S_1)$ in the complexity space are depicted schematically. As a collection of cooperative systems moves along such a trajectory its structural complexity increases monotonically.

The optimal algorithm provides the following important condition used in the computational experiments. For any TSP problem the collection of N cooperative systems proceeds along a trajectory

$$C(S_0) \prec C(S_v) \prec C(S_1), \quad 0 \le v \le 1 + \varepsilon,$$

in the complexity space. This trajectory in general depends on the TSP problem but irrespective of any problem starts in $C(S_0)$ and finishes in $C(S_1)$ when parameter v varies from 0 to $1 + \varepsilon$ (see Figure 7.3 for illustration). This allows for the optimal algorithm to control the structural complexity of the cooperative systems as it becomes an increasing function of the parameter v. In the computational experiments the focus of our interest is to understand how average performance of the cooperative systems varies with the increase of their structural complexity.

For an optimization problem there were a unique value of parameter v where the performance peaked on the interval, it could open a way to characterise the optimization problem by a corresponding complexity matrix. Then this matrix could be used as a means to define structural complexity of the problem.

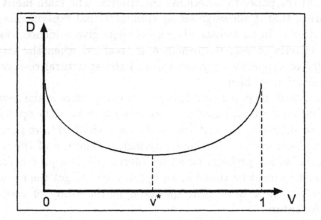

Figure 7.4. The figure schematically shows the main result of the computational experiments. For any optimization problem tested there is a point v^* where the average performance peaks as soon as the number N becomes large enough.

Extensive computational experiments by using the same class of benchmark TSP problems have been done to see how average distance traveled by the group

$$\bar{D}(v) = \frac{\sum_{i=1}^{N} D_i(v)}{N}$$

depends on v. The computational experiments have demonstrated clearly that as the number of cooperative systems N increases the function $\bar{D}(v)$ has the tendency to become an unimodal function with one global minimum. This means that for each TSP problem there is a value of parameter v such that the collection of cooperative systems as a whole demonstrates its best performance. Figure 7.4 schematically illustrates the fact that for any optimization problem tested there is a point v^* on the interval where the average performance peaks. For an optimization problem such points make possible to associate it with corresponding complexity matrices and use them to define structural complexity of the optimization problem.

Finally, this allows us to suggest that the criteria of best performance are realized when the structural complexity of cooperative systems equals the structural complexity of the optimization problem.

6.   Conclusion

A general framework based on a concept of structural complexity is proposed in the paper to determine the criteria. The main merit of this framework is that it allows to set up computational experiments revealing the criteria. In particular, the experiments give evidence to suggest that the criteria of best performance are realized when the structural complexity of cooperative systems equals the structural complexity of the optimization problem.

The results of the paper could give a new perspective in the developing of optimization methods based on cooperative systems. The optimization methods according to the criteria should be devised to have parameters controlling their structural complexity handy. The use of the methods for an optimization problem could be interpreted as a process in which the parameters must be tuned in an optimal way to get the criteria.

This raises many interesting questions in the study of cooperative systems for future.

References

[1] R. Axelrod, (1984), *The Evolution of Cooperation*, Basic Books, New York.

[2] M. Mezard, G. Parisi and M. A. Virasoro (1987), *Spin Glass Theory and Beyond*, World Scientific, Singapore.

[3] B. Huberman, (1990), *The Performance of Cooperative Processes*, Physica D, vol. 42, pp. 39-47.

[4] J. Holland, (1998), *Emergence: From Chaos to Order*, Perseus Books, Massachusetts.

[5] E. Bonabeau, M. Dorigo and G. Theraulaz, (1999), *Swarm Intelligence: From Natural to Artificial Systems*, Oxford University Press, Oxford.

[6] V. Korotkich, (1999), *A Mathematical Structure for Emergent Computation*, Kluwer Academic Publishers, Dordrecht/Boston/London.

[7] V. Korotkich, (1995), *Multicriteria Analysis in Problem Solving and Structural Complexity*, in Advances in Multicriteria Analysis, edited by P. Pardalos, Y. Siskos and C. Zopounidis, Kluwer Academic Publishers, Dordrecht, pp. 81-90.

[8] E. Prouhet, (1851), *Memoire sur Quelques Relations entre les Puissances des Nombres*, C.R. Acad. Sci., Paris, vol. 33, p. 225.

[9] A. Thue, (1906), *Uber unendliche Zeichenreihen*, Norske vid. Selsk. Skr. I. Mat. Nat. Kl. Christiana, vol. 7, p. 1.

[10] M. Morse, (1921), *Recurrent Geodesics on a Surface of Negative Curvature*, Trans. Amer. Math. Soc., vol. 22, p. 84.

[11] ftp://titan.cs.rice.edu/public/tsplib.tar

[12] S. Weinberg, (1993), *Dreams of a Final Theory*, Vintage, London.

[13] R. Monasson, R. Zecchina, S. Kirkpatrick, B. Selman and L. Troyansky, (1999), *Determining Computational Complexity from Characteristic 'Phase Transitions'*, Nature, vol. 400, pp. 133-137.

Chapter 8

COOPERATIVE MULTI-AGENT CONSTELLATION FORMATION UNDER SENSING AND COMMUNICATION CONSTRAINTS

Lit-Hsin Loo[1], Erwei Lin[1], Moshe Kam[1] & Pramod Varshney[2]

[1] *Drexel University, Data Fusion Laboratory,*
3141 Chestnut Street,
Philadelphia, PA 19104.

[2] *Syracuse University,*
Sensor Fusion Laboratory,
Syracuse, NY 13244.

Abstract A group of cooperating vehicles (smart bombs, robots) moves toward a set of (possibly moving) prioritized destinations. During their journey towards the destinations, some of the vehicles may be damaged, but it is also possible that reinforcements may arrive. The objective is to maximize the number of encounters between the vehicles and the high-priority destinations. In this preliminary study, we explore how position sensors and (possibly fading) communication channels can assist the group in performing its task. We show how joint operation, and exchange of observations and estimates, improve convergence. On the other hand unfavorable cooperation attempts can sometimes lead to oscillations and confusion. In unfavorable circumstances our agents suspend cooperation and engage in "every agent for itself" mode. One of the interesting features of the proposed architecture is that individual team members can predict the success or failure of the cooperation mechanism by testing the consistency of assigned target destinations.

Keywords: sensing, communication, cooperating vehicles

R. Murphey and P.M. Pardalos (eds.), Cooperative Control and Optimization, 143–169.
© 2002 *Kluwer Academic Publishers.*

1.   Introduction

Cooperative multi-agent control is required when a group of agents carries out shared tasks such as joint navigation, task allocation, and resource sharing. A multiple-agent team exhibits *cooperative behavior* if, given a task specified by the designer, the team uses an underlying mechanism ("the mechanism of cooperation") to achieve an increase in its total utility [3]. Under the multi-agent cooperative control paradigm, the agents can be teams of people, robots, aircraft, satellites, or other computational elements. Together, they perform tasks in a coordinated, reactive, and adaptive manner, geared towards a shared goal. Multi-agent cooperative control is useful in solving problems such as target seeking and engagement [7], wide area search munitions [11], weapon target assignment [14], and flying formations for satellites [20]. The paradigm has several advantages, including robustness, extended coverage, better resource allocation (and hence lower cost,) and increased accuracy [3, 7]. In this study, we concentrate on the cooperative creation of a formation by a group of vehicles (*e.g.*, robots, smart bombs) where the objective is to arrive at a set of (possibly time varying) locations. These locations, or targets, are assigned priorities, and in general we desire to have as many of them 'hit' by the vehicles. This objective requires dynamic allocation of targets to vehicles and continuous re-assessment of the mission. The re-assessment is based on vehicle and target status, communication constraints, and capabilities of sensors. Naturally, attention must be given to the optimal use of energy (*e.g.*, fuel and battery power).

The traditional approach to multi-agent problems of this kind has been to employ a central or hierarchical structure, making certain agents responsible for directing actions of others in the quest for achieving the common goal. In order to compute the optimal plan to be followed by each agent, central planning and explicit world representation were required. "Central" agents were assigned to monitor and oversee the activities of the other agents. However such centralized architectures are largely unsuitable for many military applications. Often these architectures are unable to respond to dynamic situations, and are not robust in the face of malfunctioning and damaged agents. An alternative approach is to organize a decentralized team of peers that are generally identical (or comprise a small number of "types"). Each agent carries out rather simple functions independently, with very limited or no communication with its peers. Architects of these structures hope that the agents can be designed to exhibit jointly an "emergent behavior" that will contribute to the common goal. The resulting agent teams are often called *swarms*, after the insect societies on which they are modeled [5, 8,

19, 22]. However, while swarms of a large number of homogeneous agents do exhibit robustness, their activities are not guaranteed to converge to an acceptable solution. Even when convergence is achieved, it often requires excessively long periods of time, making the approach undesirable for time-critical applications.

In this paper, we study an architecture consisting of autonomous homogeneous agents, who are able to communicate with their peers (though communication may be intermittent). The agents' activities and modes of operation are modified according to the conditions of the communication channels (*e.g.*, the available SNR, capacity, and "jamming status" of the channel). Each agent can carry out its tasks independently, namely it is capable of contributing to the joint task even when it has no contact with its peers. However, it does possess the means that would allow it to communicate. When the conditions allow communication, the agent would use this opportunity to modify its goals and plans, making use of the information exchanged with other agents. One of the potential advantages of communication among peers is that it may speed convergence.

Preliminary work on this architecture has been carried out by Arkin *et al.* [1,2] for robotics applications. Results were inconclusive. Arkin has shown that introducing communication improves performance in most cases, but has also found a situation that exhibited no benefits from the added communication. Additional studies in this direction were reported by researchers affiliated with the Oak Ridge National Laboratory [12, 17, 20].

2. Group formation by autonomous homogeneous agents

The group formation task that we study here is to organize a group of vehicles or target-seeking bombs in L cohorts, each having a single *leader* (physical or virtual) and a number of *followers*. The followers would maintain fixed relative positions with respect to the (moving) leader.

Different variations of the problem were discussed by Gentili *et al.*[9], Yamaguchi and Burdick [23], Yamaguchi [23], Desai and Kumar [4], and Folta *et al.*[6]. We want to use this problem to understand sensing and communication constraints in groups of cooperating target seekers, especially when the number of functioning seekers in the group changes with time due to damage inflicted by the enemy and the arrival of reinforcements.

We consider a group of $N(k)$ vehicles, where k is a discrete-time variable. The vehicles could be actual transportation platforms, or a group of robots, or a group of bombs destined to hit a set of targets. At time $k = 0$ the $N(0)$ vehicles occupy random initial positions in a certain subset of R^2. They are to form a rigid formation with respect to a (moving) frame of reference, by positioning themselves in specified positions measured with respect to a set of L leading vehicles (the "leaders"). The leaders are potentially in motion along specified paths, though operational constraints may cause them to deviate from these paths. The number of available followers may change in time (due to damage to some vehicles or the arrival of reinforcements), resulting in changes in the desired formation in mid-course. Depending on different constraints in the environment, each follower may or may not be able to sense the locations of other vehicles, and sensory data may be noisy and intermittent. Different objective functions for the group can be contemplated, such as minimum distance traveled by all followers, minimum time required to create the formation, or minimum weighted sum of the distances to the formation's positions, resulting in an ordering priority among the formation positions. The last case is important when there are some targets that are more important than others, and the objective is to hit as many of the high-priority targets as possible.

Specifically we consider here several numerical examples involving a group of $N(k) + 3$ robots, of which three are *leaders* and $N(k)$ are *followers* (k is a discrete-time variable). The leaders in our example (L_1, L_2, L_3) will be stationary, with positions (x_{L1}, y_0) (x_{L2}, y_0) (x_{L3}, y_0). Here x_{Li} is the position of the ith leader along the X-axis with $x_{L1} < x_{L2} < x_{L3}$, and y_0 is a common Y-axis position. Needless to say it is of importance to study moving leaders, but this objective, solving the *cooperative rendezvous problem*, is beyond the scope of the present study.

At t_0 the followers are scattered in (random) initial positions in the plane. Let $\lfloor x \rfloor$ be the largest integer less than or equal to x. In our example the objective is to arrange the group so that it moves in three convoys with $\left\lfloor \frac{N(k)}{2} \right\rfloor$ of the followers led by L_2 (the middle leader), $\left\lfloor \frac{N(k) - \left\lfloor \frac{N(k)}{2} \right\rfloor}{2} \right\rfloor$ led by L_1, and the rest led by L_3. The followers are to position themselves along a straight line behind the leader along the Y-axis, in positions $(x_{Li}, y_{Li} - m\Delta y)$ where m is an integer that goes from 1 to the number of followers in the convoy. We number the positions in the formation according to the priorities of filling them with followers, as illustrated in Figure 8.1 for $N = 11$. The priority pattern shown in Figure 8.1 gives preference to positions that are closer to the leader over other positions, and to the leader in the middle, L_2, over the other leaders. We assume that the maximum speed of each follower is v_m.

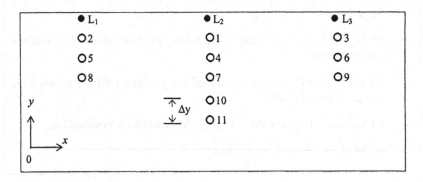

Figure 8.1. Formation Structure and priorities for a 11-vehicle convoy (lowest number ⟺ highest priority)

3. The noiseless full-information case

We consider first a simple solution to the group formation problem, assuming that the number of followers is constant $N(k) = N$, and that positions of all leaders and followers are known to all vehicles at all times. Given the initial positions of all followers, the task is to solve an Assignment problem of the form

$$\text{Minimize} \quad \sum_{i=1}^{N} \sum_{j=1}^{N} t_{ij} x_{ij} \qquad (1)$$

Subject to

$$\sum_{j=1}^{n} x_{ij} = 1 \quad i = 1, 2, \ldots, N \qquad (2)$$

$$\sum_{i=1}^{n} x_{ij} = 1 \quad j = 1, 2, \ldots, N \qquad (3)$$

$$x_{ij} = 0 \ or \ 1 \quad i = 1, 2, \ldots, N \quad j = 1, 2, \ldots, N \qquad (4)$$

Where $x_{ij} = 1$ if follower i was assigned to position j, and t_{ij} is a function of the distance between follower i and position j, d_{ij}.

P1- Minimum-distance: If $t_{ij} = d_{ij}$ we minimize the total distance traveled by all followers while creating the formation.

Table 8.1.   The basic target-seeking algorithm

Set $t = 0$

(BA1) Each follower obtains or estimates the positions of all leaders and followers.

(BA2) Each follower solves one of the problems P1 - P3 based on the performance index.

(BA3) Each follower moves by $\nu_m \Delta t$ toward its destination.

(BA4) $t \leftarrow t + \Delta t$, go to (BA1).

P2- Formation building in priority order: We can set $t_{ij} = w_j d_{ij}$ where w_j is a monotonically decreasing function in j. This will allow us to speed the filling of higher priority positions when we solve the Assignment problem.

P3- Minimum time: To create the formation in minimum time, all N! assignments should be considered, and the one chosen should have the smallest value of $\max_{i,j} d_{ij}$ over all possible assignments[1]

Table 8.1 provides a simplistic algorithm, based on geometric considerations, for positioning the followers in formation.

The task of each follower is to ascertain at each time step, that the number of followers did not change, and that the followers' positions are as expected. If a change in either is detected, the optimization problem needs to be solved under the new conditions, namely new destinations are assigned to the vehicles. Suitable precautions are needed to prevent oscillations in the assignment of destinations due to small position-measurement errors.

In this context, it is interesting to search for an alternative procedure if a probability distribution on the number of surviving followers at time k, $N(k)$, is given. One possible approach is to direct the followers toward a position that is the weighted average of all formation positions, $\hat{x}_i = \sum_{p=1}^{N} d_{ip} x_p$, where d_{ip} is the probability that the ith follower will ultimately go to the pth position. Once the follower is within a small distance of $\hat{x}_i$ it would direct itself toward its highest priority destination.

Figures 8.2-8.4 provide a performance yardstick for our system, using examples of 15-75 vehicles. The graphs show the mean time to task completion (Figure 8.2); mean total distance traveled by the vehicles, which is also an indication of fuel consumption (Figure 8.3); and the standard deviation of the total distance (Figure 8.4). The graphs are provided under four conditions: namely (1) with no disturbances or noise; (1) with disturbance (at time t=25, the location of 50 vehicles is corrected due to a previous error in reporting of a milestone), but no noise; (2) with noise (zero-mean, variance 10) but no disturbance; and (3) with noise and disturbance. The observed behavior is not surprising. Mean time to task completion is flat in the number of vehicles. Mean total distance traveled is linear in the number of vehicles. Disturbance and noise slow convergence to the solution.

4. Limitations on communications and sensing

So far we have assumed that all vehicles possess exact measurements on the locations (and hence destinations) of all other vehicles. To get a better understanding of practical limitations, we shall use a more realistic dynamic model of the ith follower. The model assumes that the inputs to the system are the horizontal and vertical velocity commands, and that additive noise is part of the model of the plant, as well as of the observation.

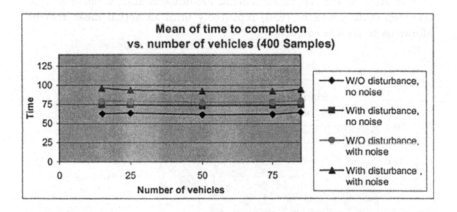

Figure 8.2. Mean time to reach targets vs. number of vehicles

Following Stillwell and Bishop [17], we shall use a decentralized control architecture where each agent is guided by a local controller that uses an estimate of the complete system state available from a locally implemented observer. Each follower will use a local Kalman filter to es-

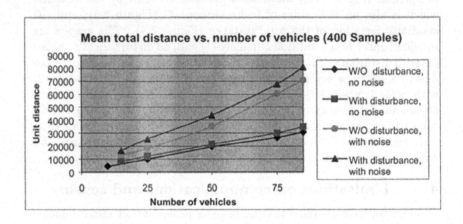

Figure 8.3. Mean total distance traveled on way to target vs. number of vehicles

timate its own position, as well as the position of all other followers. The equations of the filter are provided in Table 8.2. In this sequence, the estimation of positions may or may not be assisted by new measurements at each step. When new measurements are not available, we shall use the one step state predictor (K-6) repeatedly until an actual measurement allows us to issue a correction.

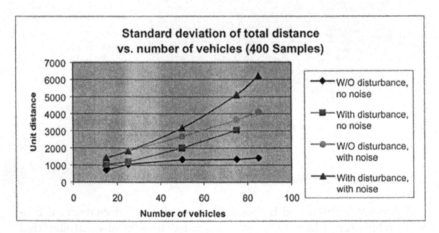

Figure 8.4. Standard deviation of total distance traveled on way to targets vs. number of vehicles

Thus the dynamics of the ith follower are:

$$
x^i(k+1) = \begin{bmatrix} x_h^i(k+1) \\ x_v^i(k+1) \end{bmatrix} = \overset{A}{\begin{bmatrix} 1 & 0 \\ 0 & 1 \end{bmatrix}} \overset{x^i(k)}{\begin{bmatrix} x_h^i(k) \\ x_v^i(k) \end{bmatrix}} +
$$
$$
\overset{B}{\begin{bmatrix} \Delta t & 0 \\ 0 & \Delta t \end{bmatrix}} \overset{u(k)}{\begin{bmatrix} v_h(k) \\ v_v(k) \end{bmatrix}} + \overset{G}{\begin{bmatrix} \sigma_h & 0 \\ 0 & \sigma_v \end{bmatrix}} \overset{w(k)}{\begin{bmatrix} w_h(k) \\ w_v(k) \end{bmatrix}} \tag{5}
$$

And the observation model (i^{th} follower measures the position of the jth follower) is

$$
z^{ij}(k+1) = \overset{C}{\begin{bmatrix} 1 & 0 \\ 0 & 1 \end{bmatrix}} x^i(k+1) + r_{ij} \overset{H}{\begin{bmatrix} \sigma_{mh} & 0 \\ 0 & \sigma_{mv} \end{bmatrix}} \begin{bmatrix} s_h(k) \\ s_v(k) \end{bmatrix} \tag{6}
$$

Example Figures 8.5 and 8.6 provide an assessment of the role that the Kalman filter (KF) plays in achieving convergence on the targets. In Figure 8.5, we show the number of the vehicles that have reached targets (versus time) for groups of 75 and 100 vehicles. In both cases, the vehicles move in a field of dimensions 1000 x 1000 (in arbitrary distance units). They experience a plant noise variance of 40 and an observation noise variance of 150. All vehicles have a constant speed of 10 units distance/unit time. We show what happens when a KF is not present and what happens when it is added. In Figure 8.6 we show the mean distance (which is proportional to fuel) needed by the system of robots to get to the destination with and without the KF (vs. the number of robots).

We note that without a filter the 100-vehicles group never gets to final formation. In fact, the vehicles continue to "wander", and are confused by the noisy messages received from their peers. Such situations are expected to occur at some level of noise and interference even in a system with KFs. One of the useful features here is that when the vehicles are unable to accomplish the formation task, they are aware of this inability (they exhibit frequent oscillations in the assigned targets, and the destination assignments do not converge). Once a vehicle detects that cooperative control does not work, it can switch to local control and move toward the more important targets irrespective of the actions of other vehicles. This "disengagement" option assures coverage of the most important targets even when noise limitations make cooperation impossible.

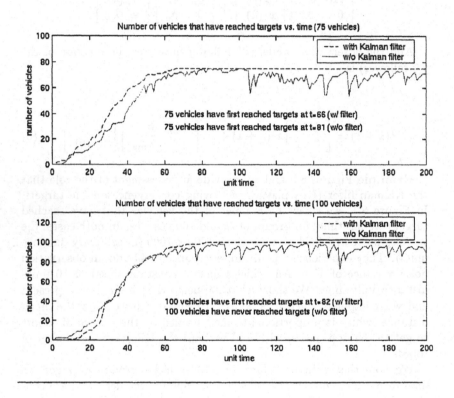

Figure 8.5. : Number of vehicles that have reached a target vs. time (for 75, 100 vehicles)

Table 8.2. Estimation of the follower's position

Plant Model	$x(j+1) = Ax(j) + Bu(j) + Gw(j)$	(M-1)
Observation Model	$z(j) = Cx(j) + Hs(j)$	(M-2)
Prior Statistics	$E\{w(j)\} = 0 \quad E\{s(k)\} = 0 \quad E\{x(0)\} = \mu_x(0)$	(K-1)
	$cov\{w(j), w(k)\} = I\delta_k(j-k)$	(K-2)
	$cov\{s(j), s(k)\} = I\delta_k(j-k)$	(K-3)
	$cov\{w(j), v(k)\} = 0$	(K-4)
	$var\{x(0)\} = V_x(0)$	(K-5)
One-stage predictor	$\hat{x}(j+1|j) = A\hat{x}(j) + Bu(j)$	(K-6)
A priori variance algorithm	$V_{\tilde{x}}(j+1, j) = AV_{\tilde{x}}(j)A^T + GG^T$	(K-7)
Filter gain algorithm	$K(j+1) = V_{\tilde{x}}(j+1|j)C^T[CV_{\tilde{x}}(j+1|j)C^T + I]^{-1}$	(K-8)
Filter algorithm	$\hat{x}(j+1) = \hat{x}(j+1|j) + K(j+1)[z(j+1) - C\hat{x}(j+1|j)]$	(K-9)
A posteriori variance algorithm	$V_{\tilde{x}}(j+1) = [I - K(j+1)C]V_{\tilde{x}}(j+1|j)$	(K-10)
Initial conditions	$\hat{x}(0) = \hat{x}(0|0) = \mu_x(0) = \varepsilon\{x(0)\}$	(K-11)
	$V_{\tilde{x}}(0) = V_{\tilde{x}}(0|0) = var\{x(0)\} = V_x(0)$	(K-12)

5.   Limitation of communications

When updating its prediction (K-6), the Kalman filter requires information about the control vector u (which in our case is the velocity input to the vehicles). This requirement introduces two difficulties: (i) the velocity inputs of a follower are not known to other followers; (ii) due to measurement errors, more than two followers may decide to proceed toward the same destination.

If no communication is available between followers, the best that the kth follower can do is to predict where the lth follower is headed, and then estimate the lth follower position $\hat{x}^l(j+1|j)$ by moving the lth follower $v_m\Delta t$ toward the predicted goal. The kth follower can do this in one of the following ways:

The kth follower solves Pi, and uses the solution to predict where all followers are headed (the kth follower may be wrong due to errors in its measurements).

The kth follower solves Pi in order to decide where it is headed, but uses for each one of the other followers the following procedure:

(B-1) Determine the N possible locations for follower k in the next step $\hat{x}_i^l(j+1|j)$ by moving it $v_m\Delta t$ toward each one of the N destinations.

(B-2) Select the location that minimizes $|z^l(j+1) - \hat{x}_i^l(j+1|j)|$.

(B-3) Calculate $\hat{x}^l(j+1)$ (equation K-7).

This approach is in the spirit of Moghaddamjoo and Kirlin [13]. We found the other formal approaches for optimal filtering with unknown inputs (*e.g.*, Hou and Patten [10] and its references) gave less satisfactory

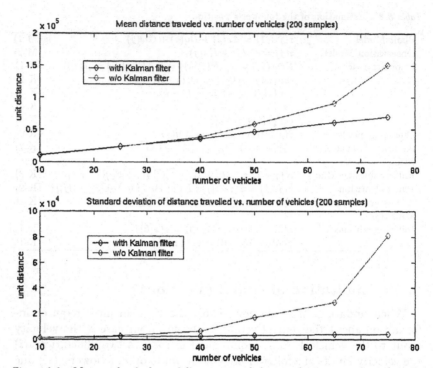

Figure 8.6. Mean and std. dev. of distance traveled to reach a target vs. number of vehicles

results, since they did not take advantage of available specific knowledge about the inputs u.

The difficulties described so far in estimating follower positions by other followers can be alleviated considerably through communication between the followers. Again, there are several ways of doing this. Followers can announce their position estimates or their own observations of positions (own positions and those of other followers). Then each follower can combine these estimates, using the usual Kalman filtering argument to get a better estimate for the location of each other follower. So, given a set of n observations of the position of a certain follower, we shall calculate

$$\hat{x} = \sum_{i=1}^{n} z_i \frac{\displaystyle\prod_{j=1, j\neq i}^{n} \sigma_j^2}{\displaystyle\sum_{k=1}^{n} \prod_{j=1, j\neq k}^{n} \sigma_j^2}, \tag{7}$$

where the observation z_i is known to be embedded in additive zero-mean Gaussian noise of variance σ_i^2. Each follower can then solve Pi and find the correct joint assignment of all followers. Since the information used in solving Pi is common in perfect communication, we shall avoid assigning more than one follower to the same destination.

Example To demonstrate the performance increase due to communication, we consider a group of 10 vehicles in a 1000 x 1000 unit-distance field with zero mean plant noise with variance (Gw in equation M-1) of 40 per axis. We assume that the observation noise (Hs in equation M-2) is zero mean, with variance that proportional directly to the distance between two vehicles.

$$\sigma_{ij}^2 = \sigma_{ii}^2 + r\left(\sqrt{(x_i - x_j)^2 + (y_i - y_j)^2}\right) \text{ for } i = 1\cdots n; \; j = 1\cdots n. \quad (8)$$

Here σ_{ij}^2 is the variance (per axis) of the position of vehicle j, measured by vehicle i. n = number of vehicles, x = x-coordinate position of a vehicle, y = y-coordinate position of a vehicle and r = a proportionality constant. σ_{ii}^2 is the observation noise for the i-th vehicle while it observes its own position. We used $r = 10$ and $\sigma_{ii}^2 = 5$.

When there is no communication between vehicles, each vehicle runs n Kalman filters, each using observations of positions of the other vehicles, governed by (8). When communication is available, each vehicle runs a single Kalman filter for itself, broadcasts its position estimate to all other vehicles, and gets estimates from other vehicles. These common estimates are then used to calculate each vehicle's destination and plan its next step. If communication is intermittent, then each vehicle runs n Kalman filters at all times, and uses the best available observation to update the filters' estimates. It either uses its sensing of other vehicles' positions, or employs the more-accurate position estimates that may be available through communications.

We assume that communication is transmitted through Rayleigh fading channels between vehicles. The distribution of the received signal, a, is therefore

$$\rho(a) = \frac{a}{\sigma^2}\exp\left[-\frac{a^2}{2\sigma^2}\right] \; for \; a \geq 0 \quad (9)$$

Our assumption is that a signal whose power is weaker than the power of the mean signal by 10dB or more, is too weak for communications. Table 8.3 shows the distance and time traveled by the 10-vehicles cohort before convergence was observed. This is done with no communication, with perfect communications, and with fading channels (using a 10dB cutoff). Communications provide a 15%-16% savings in distance and time traveled. As "more" communication becomes available, the vehicles travel less distance and require less time to converge to a solution.

Table 8.3. Time to reach target vs. radius of field of view for a single agent

Condition	Distance traveled		Time traveled	
	Mean	St. Dev.	Mean	St. Dev.
1. Without communication	11574.0	1235.3	98.3	11.21
2. With Rayleigh fading channel (cut-off at −10dB with respect to mean)	9848.3	440.6	83.0	4.11
3. With perfect communication	9775.6	472.0	82.3	4.38

6.　　Oscillations due to sensing limitation

When the observation noise is high, it is possible that more than two agents will be headed towards the same target. We repeat the last example with $r = 15$ and $\sigma_{ii}^2 = 5$. The trajectories of the vehicles are shown at t=65 without communication (Figure 8.7) and with intermittent communication (Figure 8.8). In Figure 8.7, we note that two vehicles have been wandering around target 5 for a long time. The situation was much improved in Figure 8.8. The same two vehicles wandered about target 5 for a while, but with assistance of communications, one was able to "escape" and continue towards target 9.

When noise levels are high, especially without communications, vehicles would have very poor estimates of the positions of other vehicles. Assuming that (8) is an appropriate model for the variance, remote vehicle will be practically unobservable. Vehicles will then ignore some or most other vehicles, move toward the highest priority target, and then re-examine the situation. At close proximity to the highest priority target, a vehicle may either find itself alone (and therefore continue to pursue the target) or at close proximity to several other vehicles. Being able to see other vehicles better at shorter distances, the vehicle may now reassign itself to lower priority targets. Unfortunately targets can get "in-and-out" the field of view, and it is then possible for the vehicle to experience frequent re-assignments of targets and oscillations in its destinations. A preliminary analysis of this situation appears in the next section.

7.　　Group formation with partial view

So far, we have assumed that all vehicles are able to measure the locations of other vehicles at all times, albeit with less accuracy when

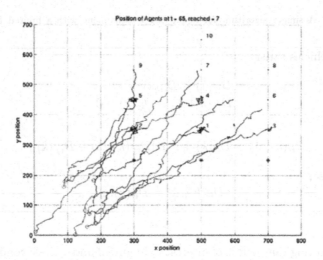

Figure 8.7. Position of agents at t=65 without communication

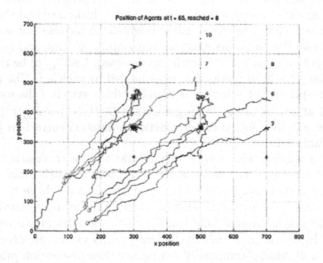

Figure 8.8. Position of agents at t=65 with communication

distances between vehicles are large. In practice, the range of view of each vehicle is limited by the large variance of remote vehicles, and by

Table 8.4. A simple algorithm for agent-target assignment with a limited field of view

If any vehicles satisfy

$$\hat{D}_{ij}(k) < \hat{D}_{ii}(k)$$
$$\hat{D}_{jj}(k+1) - \hat{D}_{ij}(k) > 0$$

then

$$\hat{D}_{ii}(k+1) = \hat{D}_{ii}(k) + \max_{j}[\hat{D}_{ij}(k+1) - \hat{D}_{ij}(k)].$$

Otherwise,

$$\hat{D}_{ii}(k+1) = \hat{D}_{ii}(k).$$

limited sensing range. It is of interest to inquire if under these conditions purposeful and efficient team navigation toward the goals is still possible.

When the ith vehicle sees a number $N_i(k) < N(k)$ of vehicles at time k, it may assume first that it sees the complete field, and therefore assign all vehicles in sight, including itself, to the high-priority destinations. However, as early as one unit of time later the vehicle may find that the actual one-step progress of its visible neighbors is inconsistent with its assumptions. The neighbors must be seeing other vehicles, and may not see some of those that the ith vehicle has detected. Let $\hat{D}_{ij}(k)$ be the priority of the jth vehicle destination as estimated by vehicle i. The actual destination-priority of vehicle j is $\hat{D}_{jj}(k)$, which vehicle i discovers at state $k+1$ after measuring j^{th} progress (thus $\hat{D}_{ij}(k+1) = \hat{D}_{jj}(k)$). The ith vehicle will then follow the destination-estimation correction procedure of Table 8.4.

Consider a graph with a node at the location of each vehicle. Let a branch exist between two nodes if the two robots see each other. Each R-node connected component of this graph assigns a vehicle to the R most important destinations. Ideally, as the vehicles move toward the target, the graph becomes fully connected; the vehicles are sorted in a manner which guarantees that the highest-priority targets are covered at all times, and vehicles continually downgrade their destination priorities as they see more vehicles. Success depends on all vehicles being able to re-orient themselves to the optimal target before impact.

Figures 8.9 and 8.10 demonstrate how ten vehicles adjust their priorities as they approach ten targets. After 10 steps many of the vehicles still do not see any neighbors, and consider themselves to seek the priority-1 target. As they get closer to the common-target area (Figure 8.10)

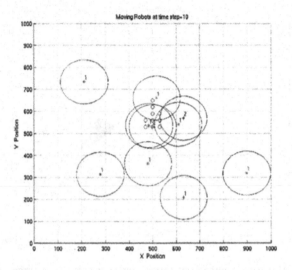

Figure 8.9. Ten vehicles approaching a target cluster, $t = 10$

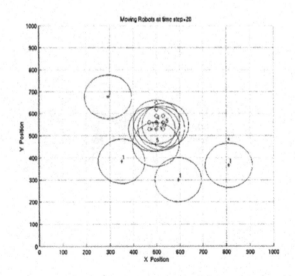

Figure 8.10. Ten vehicles approaching a target cluster, $t = 20$

they readjust their destination, relinquishing the higher-priority targets
to vehicles which are closer to these targets.

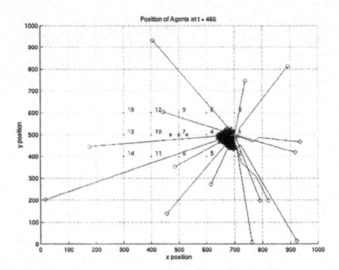

Figure 8.11. 15 agents approaching target cluster without new target assignment scheme at $t = 480$

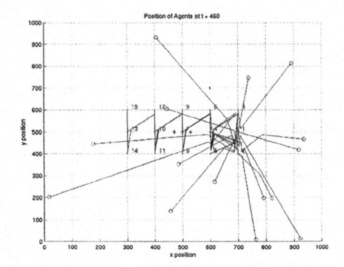

Figure 8.12. 15 agents approaching target cluster with new target assignment scheme at t=480

We concede that this preliminary algorithm (Table 8.4) is prone to oscillations and unnecessary changes of course, due to intermittent appearance of neighboring vehicles in sight. In order to avoid oscillations, we can use a monotonically decreasing priority target assignment scheme. Under this scheme, each agent always starts by assigning itself to the highest priority target. If agent A detects another agent B, which is closer to agent A's destination, then agent A assigns itself to the next-highest priority target, and changes its course towards this new destination. Agent A will not be assigned to a higher priority destination unless it is able to detect the loss of agent B, who is headed toward a higher-priority target, while agent B is still within agent A's sight of view. By forcing agent A to always downgrade its destination priority, we are able to avoid oscillations. The price is that we are unable to account for losses of targets due to agent expiration outside the field of view of active agents. If a high-priority target was to be attacked by an agent that was destroyed "out of sight", its target won't be assigned to a new agent automatically.

Example Consider 15 agents moving at constant speed of 5 unit distances/unit time with radius of sight of 20 unit distances on a 1000x1000 unit-distance field. There is no communication and observations are noise-free. Figure 8.11 shows the oscillating behavior of the original algorithm (Table 8.4). All agents wandered around target 1 which has the highest priority. The problem was solved in Figure 8.12 with the monotonically decreasing priority target assignment scheme. All targets were successfully covered in spite of the limited field of view of the agents.

The relationship between the mean total time to completion and the radius of sight of an agent is shown in Figure 8.13 (average of 100 runs with initial positions of 15 agents distributed randomly over a 1000x1000 field). The time needed for completion decreases significantly as the field of view becomes larger.

8. The use of 'meeting point' for target assignment

Although the monotonically decreasing priority target assignment scheme can solve the online assignment problem, there may be better schemes, even if no agent loss may occur. One of the possible deficiencies of our original algorithm becomes evident as agent get close to the targets. They may discover many competitors not seen before, necessitating big (and wasteful) changes in their plans. It may be useful to direct the agents to congregate first at a central location, and decide their final destinations. This is particularly useful if the times needed to reach the meeting point are roughly the same for all agents. Using this method, all agents will move directly to central point before they head to

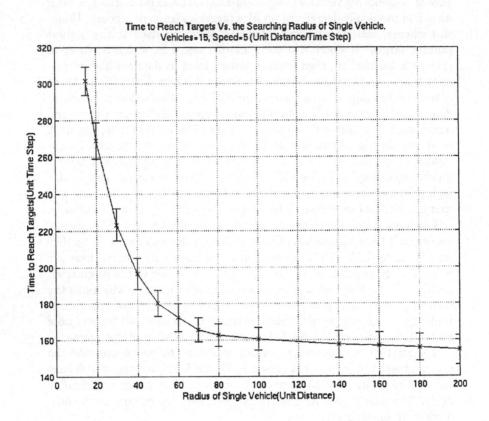

Figure 8.13. Time to reach target vs. radius of field of view for a single agent (mean and standard deviation)

the real targets. The central point of the cluster of target is its centroid (X_C, Y_C), which satisfies

$$Min_{(X_C,Y_C)} \sum_{i=1}^{N} \sqrt{(X_{T,i} - X_C)^2 + (Y_{T,i} - Y_C)^2} \qquad (10)$$

where $(X_{T,i}, Y_{T,i})$ is the location of the i-th target.

The minimization in (10) is an unconstrained optimization problem in two variables, and there exits several ways to solve it. We choose an iterative gradient search-based algorithm [15] to find (X_C, Y_C) because the algorithm is more robust and requires less calculation compared with other nonlinear optimization algorithms.

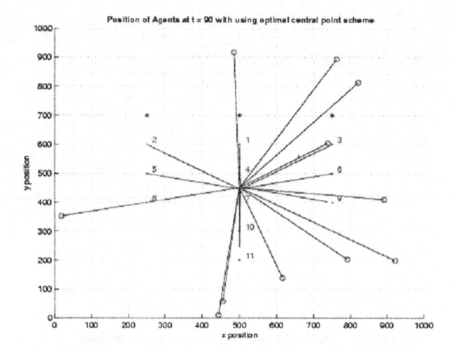

Figure 8.14. Position of Agents at t=90 by using optimal central point scheme.

Figure 8.14 shows the centroid of 11 targets and 11 agents approaching, each having a radius of sight of 150 units. The decision where to go next is made by the agents as they get close to the centroid. Figure 8.15 shows the improvement in the mean distance traveled obtained by meeting at the centroid, compared with using the monotonically decreasing priority assignment scheme. The improvement in this case is tied to the fact that most agents will arrive at the centroid at about the same time, though as long as all targets are seen from the centroid location this assumption can be relaxed.

We note that sending all agents to a central point may be highly undesirable in military scenarios, and in this case a more sophisticated "coordination on the fly" method may be needed.

9.   Conclusion

We provide a group of agents with means to achieve a formation in noisy environments. The main tool is a bank of Kalman filters assisted

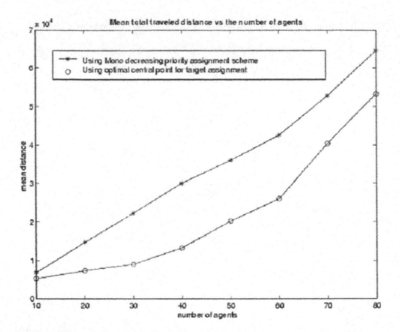

Figure 8.15.   Mean total travel distance vs. the number of agents

by potentially-intermittent Rayleigh fading communication channels. A decentralized assignment algorithm readjusts agent destination as conditions change.

Although this group formation problem shows tractability and computational feasibility, the models that we have used (two dimensions, control through velocity commands) are simplistic. There is a need to expand the study to higher order systems, with possible coupling between plant and observation noises, and consideration of nonlinear effects (leading to the use of an Extended Kalman Filter).

Dynamic allocation of targets need to be re-addressed, taking into account the expected distribution of damage to smart bombs/vehicles, making target assignments more robust, and developing algorithms for recognizing situations which do not allow for cooperative control. In general we need to make the cohort (1) more tolerant to failure of its components than the preliminary schemes presented here; and (1) more robust in the case of limited visibility.

The controllability and observability of a cooperative cohort need to be defined and used. Most important in this context is the detection of lack of controllability, when the cohort recognizes that cooperative control is impossible to execute. Under these circumstances, the system should switch to an "every one for itself" mode, which seeks to cover the most important targets in spite of the inability to coordinate operations.

A (much) more detailed study of the effect of communications on performance is needed. This study would include the effect of communication gaps of position estimation, the possibility of triangulation through fusion of neighbor measurements, and the fusion of GPS measurements and local observations. Communication is, of course, an integral part of the controllability and observability assessments. In this context, we need to address what happens when communication is possible only within a local neighborhood, and with capacity limitations.

References

[1] Arkin, R.C. et al. "Communication of Behavioral State in Multi-Agent Retrieval Tasks", *Proceedings of the IEEE International Conference on Robotics and Automation*, 1993, pp. 588-594.

[2] Balch, T. and Arkin,R. "Communication in reactive multiagent robotic systems," *Autonomous Robots*, Vol. 1, no. 1, pp. 1-25, 1994.7

[3] Cao, Y.U., Fukuaga, A., and Kahng, A. "Cooperative Mobile Robotics: Antecedent and Directions," *Autonomous Robots*, Vol. 4, pp. 7-27, 1997.

[4] Desai, J.P. and Kumar, V. "Nonholonomic Motion Planning for Multiple Mobile Manipulators", Proceedings of the IEEE International Conference on Robotics and Automation, pp. 3409-3414, Albuquerque, New Mexico, April 1997.

[5] Dorigo, M. and Di Caro, G. "Ant Colony Optimization: A New Meta-Heuristic", Evolutionary Computation, 1999. Proceedings of the 1999 Congress – 1477 Vol. 2.

[6] Folta, D., Newman, L.K. and Quinn, D. "Design and Implementation of Satellite Formations and Constellations", NASA Technical Report, Greenbelt, MD, 1998.

[7] Freilinger, D., Kvitky, J., and Stanley, W. *Proliferated Autonomous Weapons, an Example of Cooperative Behavior.* Technical Report, Project Air Force, RAND, 1999.

[8] Gelenbe, E. et al. "Autonomous search by robots and animals: A survey", Robotics and Autonomous Systems, Vol. 22, pp. 22-34, 1997.

[9] Gentili, F. and Martinelli, F. "Robot group formations: a dynamic programming approach for a shortest path computation". Proceedings of the 2000 IEEE International Conference on Robotics & Automation, San Francisco, CA, April 2000.

[10] Hou, M. and Patton, M.J. "Optimal Filtering for Systems with Unknown Inputs", *IEEE Transactions on Automatic Control*, Vol. 43, No. 3, pp. 445-449, March 1998.

[11] Jacques, D.R., Leblanc, R. "Effectiveness Analysis for Wide Area Search Munitions," American Institute of Aeronautics and Astronautics, 8pp., available on line at http://er4www.org.ohio-state.edu/~passino/uaav_background.html

[12] Jung, D. and Zelinsky, A. "Grounded Symbolic Communication Between Heterogeneous Cooperating Robots." *Autonomous Robots*, Vol. 8, No. 3, 2000.

[13] Moghaddamjoo, A. and Kirlin, R.L. "Robust Adaptive Kalman Filtering with Unknown Inputs", *IEEE Transactions on Acoustics, Speech, and Signal Processing*, Vol. 37, No. 8, pp. 1166-1175, August 1989.

[14] Murphey, R.A. "An Approximate Algorithm for a Weapon Target Assignment Stochastic Program." In *Approximation and Complexity in Numerical Optimization: Continuous and Discrete Problems* (P.M. Pardalos, editor), Kluwer Academic, 1999.

[15] Reklaitis, G.V. *Engineering Optimzation Methods and Applications.* Wiley-Interscience, 1983.

[16] Parker, L.E. "Life-long Adaptation in Heterogeneous Multi-robot Teams," *Autonomous Robots*, Vol. 8, No. 3, 2000.

[17] Stilwell, D.J. and Bishop, B.E. "A Framework for Decentralized Control of Autonomous Vehicles". *Proceedings of the 2000 IEEE International Conference on Robotics and Automation*, San Francisco, CA; April 2000; pp. 2358-2363.

[18] Sugawara, K. et al. "Foraging Behavior of Multi-robot System and Emergence of Swarm Intelligence." *Proceedings of the IEEE International Conference on Systems, Man, and Cybernetics*, 1999,. pp. 257-262.

[19] Touzet, C. "Robot Awareness in Cooperative Mobile Robot Learning." *Autonomous Robots*, Vol. 8, No. 1, pp. 87-97.

[20] US Air Force Rome Laboratory: *TechSat21 program description*, on-line: www.vs.afrl.af.mil/TechProgs/TechSat21, accessed January 2, 2001.

[21] White, T. and Pagurek, B. "Towards Multi-Swarm Problem Solving in Networks", Proceedings of *Multi Agent Systems*, pp. 333-340, 1998.

[22] Yamaguchi, H. "Adaptive Formation Control for Distributed Autonomous Mobile robot Groups." *Proceedings of the IEEE International Conference on Robotics and Automation*, pp. 2300-2305, Albuquerque, New Mexico, April 1997.

[23] Y Yamaguchi, H. and Burdick, J.W. "Time Varying Feedback Control for Nonholonomic Mobile Robots Forming Group Formations." *Proceedings of the 37th IEEE International Conference on Decision and Control*, pp. 4156-4163, Tampa, Florida, December 1998.

Chapter 9

AN INTRODUCTION TO COLLECTIVE AND COOPERATIVE SYSTEMS

Robert Murphey

Air Force Research Laboratory, Munitions Directorate
101 W. Eglin Blvd., Ste. 330, Eglin AFB, FL 32542

murphey@eglin.af.mil

Abstract   Cooperative systems are introduced to the reader as a part of a broader class of *collective* systems. A taxonomy of collective systems is defined such that each class within the taxonomy is rigorously defined based upon the mathematical constructs of team theory. It is shown that this taxonomy leads to a precise definition of cooperation and clearly separates intentional cooperation from serendipitous complementary behavior. Concepts of precedence, hierarchy, and supervision are made clear in the presence of information such that team theory and decentralized control theory are generalized into the single framework of collective systems. It is anticipated that this framework will lead to a consistent representation of cooperation in future research and new methods for solving the hard problem of non nested information structures in team theory.

Keywords: team theory, decentralized control, cooperative control, hierarchies.

Introduction

Roughly speaking, a cooperative system is defined to be a collection of dynamical entities that share information to accomplish a common, though perhaps not singular, objective. Examples of cooperative systems might include; robots operating within a manufacturing cell, unmanned aircraft in search and rescue operations or military surveillance and attack missions, arrays of micro satellites that form a distributed large aperture radar, employees operating within an organization, and software agents. The term entity is most often associated with mechanical devices capable of physical motion such as robots, automobiles, ships, and air-

R. Murphey and P.M. Pardalos (eds.), Cooperative Control and Optimization, 171–197.
© 2002 *Kluwer Academic Publishers.*

craft. However, the definition extends to include any device, person, or mathematical abstraction that exhibits a time dependant behavior.

Critical to cooperation is communication of one entity's state to another. Since the differences between communication and sensing are often blurred, these concepts must first be clearly distinguished before a mathematical definition of cooperation is developed, which is the main goal of this work. Such rigor is necessary if we wish to separate the case of serendipitous group dynamics from that of intentional cooperation.

The weakness of natural language is that through its use, many slightly different notions tend to generalize to the same term. So it is with the word "cooperative," hence the result that cooperation means something different to each person. Conversely, so many words are used interchangeably with "cooperation"; collective, collaborative, decentralized, emergent, team-based, just to name a few. In Section 2 the entire notion of cooperation will be extracted from team theory and then built upon a new generalized framework defined as *collective systems*. This framework will distinguish, mathematically, the cases of *collaboration*, which is a weaker condition than cooperation, and *complementation* which is similar to cooperation but does not require communication. Some basic results from team theory will be supplied in Section 1 to provide a springboard for this new development.

Cooperation always has some common purpose, but the individual may have other personal and perhaps conflicting objectives. Individual conflict may also be the result of being a member of other cooperating caucuses. The result is a system of cooperating systems, which implies that cooperation may assume hierarchical forms as well. In Section 3 the role of hierarchy, precedence, and supervision in cooperative systems will be explored, being careful to distinguish between lines of authority and information.

1.   Preliminaries in game and team theory

Without question, the closest theory we have for what are commonly understood to be cooperative systems is that which describes teams. Team theory was first presented by Marschak [13] in terms of a collection of decision makers, each deciding about something unique but participating in a common payoff. Marschak's conceptualization does not explicitly model communication between team members, even though some a priori recognition of the common objective is implied. Marschak's formulation does not permit conflicts of interest between team members. At its essence, Marschak's model considers N decision makers, which we will occasionally abbreviate as DMs, and a N-tuple of decisions $u = (u_1, u_2, \ldots, u_N)$ such that individual decisions are defined within some set of admissible decisions; $u_i : u_i \in U_i, i = 1, 2, \ldots, N$ with $u \in U \subset U_1 \times U_2 \times \ldots U_N$. The payoff is a function of u and the "state of the world," which is assumed to be random and unknown.

Encouraged by Marschak's work, Radner [20] studied the influence of information regarding the state of the world on the solutions of team problems. He begins with the decision vector u and state of the world, $\xi = (\xi_1, \xi_2, \ldots, \xi_N)$. Each decision u_i is assumed to be made with information regarding the state of the world,

$$z_i = H_i \xi, \tag{1}$$

thus $u_i = \gamma_i(z_i)$, $\gamma_i \in \Gamma_i$, where Γ_i is the admissible set of decision rules or control laws for DM i. The payoff is a real number and a function of u and ξ: $J(u, \xi)$. A special case results when ξ is joint normal with $E[\xi] = 0$ and covariance Σ and J is a quadratic function of the team decision u:

$$J(u, \xi) = \frac{1}{2} u^T Q u + u^T S \xi + u^T c \tag{2}$$

where S and c are a known matrix and vector respectively, and Q is positive definite to ensure J has a unique minimum in u. Since ξ is random, we are truly interested in minimizing the expected value of J as

$$\mathbf{J} = E[J(u, \xi)]. \tag{3}$$

By taking the derivative of $\mathbf{J}$ w.r.t. u_i, a necessary and, as it turns out, sufficient condition may be obtained with a solution in the form of

$$u_i = A_i z_i + b_i \tag{4}$$

$$b = -Q^{-1} c \tag{5}$$

$$Q_i A e \Sigma H_i^T = -S_i \Sigma H_i^T \tag{6}$$

where Q_i is the ith block-row of Q, e is a column vector of ones, and, as stated earlier, Σ is the covariance of ξ. The important result here is that the solution (4) is clearly linear in the information. This means that

if the information structure is static, the payoff quadratic, and the information joint zero mean Gaussian, we may limit our search for optimal solutions to the affine space.

The derivation sketched here may be found in Ho and Chu [8] who not only reproduced Radner's result, but expanded it so that the information available to a decision maker may include knowledge of other decision makers' results. That is,

$$z_i = H_i \xi + \sum_{j=1}^{N} D_{ij} u_j, i = 1, 2, \ldots, N \tag{7}$$

where a coupling between decision makers has been added in the form of D_{ij}. It is assumed that H_i and D_{ij} are matrices that all DMs have knowledge of. To preserve causality and prevent decision dependence upon itself, we require that

$$D_{ij} \neq 0 \longrightarrow D_{ji} = 0.$$

The pattern resulting from the coupling between decision makers is often referred to as an *information structure*. The coupling may be direct, in D_{ij}, or indirect; inferred by a precedent's information being embedded within $H_i\xi$. To make the information structure clear, Ho and Chu define *precedence diagrams* as follows. Consider a graph with N vertices, each vertex uniquely corresponding to a decision maker. Construct a solid directed edge between vertex v_i and v_j if $D_{ji} \neq 0$. In this case, v_i is said to precede v_j. If there exists an ordered sequence $\{r, s, t, \ldots, j, k\}$ such that r precedes s, s precedes t, $\ldots$ j precedes k, then r is said to precede k. Now consider two vertices, v_i and v_j such that i is the precedent of j. If z_i is *embedded* in z_j, that is z_j contains z_i, then a dotted directed edge is constructed on (v_i, v_j) indicating that z_i is known at j even if $D_{ji} = 0$. The dotted line is referred to as *memory-communication*. Figure 9.1 gives a simple example of a precedence diagram for the system described by equations 8- 11.

$$z_1 = H_1\xi \qquad\qquad\qquad\qquad (8)$$

$$z_2 = H_2\xi + D_{21}u_1 \qquad\qquad\quad (9)$$

$$z_3 = H_3\xi + D_{31}u_1 \qquad\qquad\quad (10)$$

$$z_4 = \begin{bmatrix} H_4\xi + D_{43}u_3 \\ z_3 \end{bmatrix} \qquad\quad (11)$$

Notice that the diagram remains the same even if $D_{43} = 0$ due to z_3 being embedded in z_4. The *discrete-time stochastic control problem* is an example where memory-communication plays a role. This problem will be discussed next.

In the parlance of game theory, Radner's team model is said to be *static*, because each decision $u_i = \gamma_i(z_i)$, where $z_i = H_i\xi$ is made independently and consequently, the order or timing of decisions is not important. The precedence diagram for a static information pattern is simply N vertices with no edges.

When precedence is added, the information pattern is said to be *dynamic* because the order of decisions must be specified to play the game. Dynamic information patterns are typically associated with games in *extensive form* [1]. There are many interesting special cases of dynamic information patterns, most notably the *classical* and *partially nested* [8, 9].

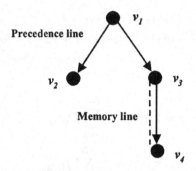

Figure 9.1.   Example Precedence Diagram

These are interesting because they may be expressed as having static information patterns as will now be shown. Consider the discrete-time stochastic control problem

$$x_{i+1} = F_i x_i + G_i u_i + w_i, \quad i = 1, 2, \ldots, N \tag{12}$$
$$y_i = H_i x_i + v_i \tag{13}$$

where x_i is the system state at stage i, y_i is the observation of the state at stage i, w_i, v_i, and x_1 are independently and normally distributed random variables (or vectors as the case may be) and F_i, G_i, and H_i are known constants (matrices). The stochastic control problem in this formulation is equivalent to a team decision problem with N decision makers such that the decision u_i at stage i is made by DM i. The information available to DM i is assumed to be the observations y_j, $j = 1, 2, \ldots, i$ and the previous stage decisions u_k, $k = 1, 2, \ldots, i - 1$, thus

$$z_i = [y_1, y_2, \ldots, y_i, \ u_1, u_2, \ldots, u_{i-1}]$$

By substituting in equations (12) and (13)

$$z_i = [H_j x_j + v_j, \ u_1, u_2, \ldots, u_{i-1}], \ j = 1, \ldots, i$$
$$= [(H_1 x_1 + v_1), H_j(F_{j-1} x_{j-1} + G_{j-1} u_{j-1} + w_{j-1}) + v_j,$$
$$u_1, u_2, \ldots, u_{i-1}], \ j = 2, \ldots, i$$

$$\vdots \qquad\qquad\qquad \vdots$$

$$= \hat{H}_i[x_1, v_1, w_{j-1}, v_j, \ j = 2, \ldots, i] + \sum_{j=1}^{i-1} D_{ij} u_j$$

$$= \hat{\hat{H}}_i[x_1, v_1, w_{j-1}, v_j, \ j = 2, \ldots, N] + \sum_{j=1}^{N} D_{ij} u_j$$

$$= \hat{\hat{H}}_i \xi + \sum_{j=1}^{N} D_{ij} u_j$$

which is in the standard form of equation (7). The dotted lines indicate a recursion on j. It is fairly obvious that since

$$z_i = [y_1, y_2, \ldots, y_i, \ u_1, u_2, \ldots, u_{i-1}],$$

then $D_{ij} \neq 0$, $\forall j < i$. Solid edges on the precedence graph from v_{i-1} to $v_i, i = 1, 2, \ldots, N$, are sufficient to illustrate the coupling. What's more,

$$z_{i+1} = [y_1, y_2, \ldots, y_{i+1}, \ u_1, u_2, \ldots, u_i]$$
$$= [z_i, y_{i+1}, u_i],$$

hence z_i is embedded in z_{i+1} for $i = 1, 2, \ldots, N-1$. This is indicated by memory-communication edges from v_i to v_{i+1}, $i = 1, 2, \ldots, N-1$. The information pattern is as shown in Figure 9.2 and is termed *classical*.

Since z_i is embedded in z_{i+1}, the term $\sum_{j=1}^{N} D_{ij} u_j$ is redundant and may be dropped. Hence, $z_i = \hat{\hat{H}}_i \xi$ which is that of the static pattern. Consequently,

> the optimal team strategy for quadratic Gaussian systems with the classical information structure is affine in the information as in equation (4).

Another, seemingly more complex, dynamic information structure is also expressible as static. The *partially nested* structure is defined by the embedding process observed in the classical pattern. Specifically, if j is precedent to i implies that z_j is embedded in z_i, then the information

Figure 9.2. Classical Information Structure

structure is partially nested. Essentially, partially nested systems have infinite memory to all precedents. The example in Figure 9.1 is not partially nested, whereas the classical pattern in Figure 9.2 clearly is. As with the classical pattern, the coupling term $\sum_{j=1}^{N} D_{ij} u_j$ in equation (7) is redundant and may be dropped, resulting in a static information pattern. Consequently,

> **the optimal team strategy for quadratic Gaussian systems with the partially nested information structure is linear in the information.**

Indeed, the partially nested structure generalizes the classical pattern. The partially nested pattern may itself be further generalized and still maintain the static equivalence as long as there exists in the precedence graph only trees, each with a partially nested structure [9].

Now consider systems without the nice embedding properties reflected by partially nested information structures. Witsenhausen [26] makes it clear that without the partially nested information structure, an affine solution can not be guaranteed. Indeed, the optimal solution may not even be obtainable, as is still the case today with the counterexample he presented over three decades ago! The counter example he presented is the simple system described by

$$x_1 \sim N(0, \sigma^2)$$
$$x_2 = x_1 + u_1$$
$$x_3 = x_2 + u_2$$
$$z_1 = x_1$$
$$z_2 = x_2 + v$$
$$u_1 = \gamma(z_1)$$
$$u_2 = \gamma(z_2)$$

with $v \sim N(0, \sigma^2)$ and $J(u_1, u_2) = k^2 u_1^2 + x_2^2$, $k^2 > 0$.

Since in z_2 we cannot distinguish x_2 from v, z_1 is not embedded in z_2. Thus we know that the system is not partially nested. Suppose we search for a solution among the affine class of decision rules, γ_1, γ_2 as suggested by equation (4). Witsenhausen gives the unique solutions as

$$u_1 = A_1 z_1$$
$$u_2 = A_2 z_2$$

For $k^2 < \frac{1}{4}$: $A_1 = A_2 - 1$, $\quad A_2 = \frac{1}{2}\left(1 \pm \sqrt{1 - 4k^2}\right)$,

For $k^2 \geq \frac{1}{4}$: $A_1 = \frac{t}{\sigma} - 1$, $\quad A_2 = \frac{\sigma^2 \frac{t^2}{\sigma}}{1 + t^2}$

where t is the unique solution of the quintic polynomial

$$(t - \sigma)\left(1 + t^2\right)^2 + \frac{t}{k^2} = 0$$

which, for $k^2 < \frac{1}{4}$ yields $J_a^* = 1 - k^2$, where J_a^* denotes the affine optimal cost. A nonlinear solution is proposed by Witsenhausen of the form

$$u_1 = \sigma \text{sgn}(z_1) - z_1$$
$$u_2 = \sigma \arctan(\sigma z_2)$$

where sgn denotes the signum function. For this formulation, the optimal cost is $J^* \to 0.404230878$ as $k \to 0$. Compare this with the cost of the affine solution for which J_a^* approaches 1 as $k \to 0$. Clearly, a nonlinear solution is superior to the best affine solution.

1.1. Non Partially Nested Information Structures.

Are non partially nested information structures common in real systems? Unfortunately, the answer is overwhelmingly yes. The most studied examples are teams of people in managerial organizations. Other examples, becoming more common place every day, are the ever increasing number of large "systems of systems" in human, power, telecommunications, entertainment, and utility networking. These networks often evolve as separate entities that are then loosely coupled to expand service coverage or as a result of a merger. Very little is understood about the quality of service or failure modes of these inter netted systems and yet our society becomes more reliant on these networks with each passing day [5, 11, 12, 16, 17, 18, 19, 23, 24, 25].

Nevertheless, as seen in the Witsenhausen counterexample, these systems are quite difficult to analyze. This does not mean that we are condemned to failure. In [6] Geanakoplos and Milgrom point out that the Marschak and Radner model discussed thus far [13, 20, 14] assumes a very limited coupling, in that the sole function of communication between decision makers is to pass along decisions and, possibly, states. Geanakoplos and Milgrom propose to add the ability for a precedent decision maker to restrict the feasible set and to supply a payoff function to another, subordinate decision maker. They argue that these capabilites are not needed in the Marschak and Radner model and therefore do not appear there since, with infinite memory and processing power, the subordinate decision maker could infer such changes for himself. Without infinite memory, one could imagine a precedent decision maker might wish to supply an *artificial objective* to a subordinate to account for the subordinate's lack of information regarding the state of the world, thereby ensuring the best collective solution. Such an example is now provided.

Example 1 (Using Artificial Objectives.) *Consider a system of 2 decision makers deciding about the level of production at 3 shops. Decision maker M allocates production levels x_1 and x_2 at shops 1 and 2 respectively while decision maker T allocates production levels x_3 and x_M to shop 3 and decision maker M. Thus T is clearly a precedent to M. The total production cannot exceed x_T. Now suppose that the production costs at shops 1,2, and 3 are respectively*

$$\frac{1}{2}(x_1 - \gamma)^2, \ \frac{1}{2}x_2^2, \ and \ \frac{1}{2}x_3^2$$

where γ is a random variable. Thus we wish to solve

$$\min \frac{1}{2}(x_1 - \gamma)^2 + \frac{1}{2}x_2^2 + \frac{1}{2}x_3^2$$
$$s.t. \quad x_1 + x_2 + x_3 \leq x_T$$

If both decision makers know the problem structure, x_T and γ, then the optimal solution is easily obtained by both as

$$x_1^* = \frac{x_T + 2\gamma}{3}, \quad x_2^* = x_3^* = \frac{x_T - \gamma}{3}.$$

If only T has knowledge of γ, then M may infer the value of $\gamma = 3x_M - 2x_T$ since he receives $x_M^* = \frac{2x_T + \gamma}{3}$ from T. Now suppose that M is unable to infer γ due to finite attention or some other inability. Then decision maker T may ensure the solution remains optimal if he supplies M with the following artificial objective

$$\min \ (x_1 + 2x_2 - x_T)^2$$
$$s.t. \quad x_1 + x_2 = x_M$$

While this example of Geanakoplos and Milgrom seems a bit contrived, it does drive home the point that there are methods to circumvent a lack of nested information. The difficulty may be in developing an artificial objective, and the success will likely be highly dependent upon the specific problem. Nevertheless, this should prove an interesting area for further research.

Without the restrictive infinite memory assumption, the Marschak and Radner model as discussed is not very useful. As the example indicates, modifications can be made to accommodate artificial objectives. Alternatively, a precedent decision maker may wish to supply artificial constraints or perhaps parameter values of the subordinate's objectives and constraints, thereby "tuning" the subordinate's response. Each of these *heteronomous modeling* concepts holds great promise as an alternative to the Marschak and Radner concept of precedence if we are to overcome the limitations of finite memory systems. Before this modification is attempted, a more precise description of collective systems is needed.

2.   Collective systems.

In this section a mathematical model for a collection of systems will be defined that will clearly distinguish the condition of cooperation from similar behaviors that may seem on the surface to be cooperation, but when examined more closely, are not. Based upon team theory, we may begin to address a fundamental question: what really constitutes cooperation? To get to the heart of this question, a broader understanding of collective systems is needed. The fact that so many variants of collective systems exist, it seems appropriate to first create a taxonomy with which to relate them. Our taxonomy will be defined with respect to the behavior exhibited by the collection. It seems logical to begin with the definition of a collective system.

Definition 1 *A* **collective** *system is a collection of at least 2 dynamical entities, henceforth termed decision makers, each with a nonempty set of individually admissible decision policies. Every decision policy may be a function of observations of the collective system state and necessarily modifies the decision maker's state in a nontrivial fashion in finite time. Mathematically,*

$$x(t) = [x_1(t) \ldots, x_N(t)], \; u(t) = [u_1(t), \ldots, u_N(t)],$$
$$\dot{x}_i(t) = f_i(x_i(t), u_i(t), t), \; z_i(t) = h_i(x(t), t), \; i = 1, \ldots, N,$$
$$N \geq 2$$

Condition 1: $u_i(t) = \gamma_i(z_i(t), t) | \; \gamma_i \in \Gamma_i \neq \emptyset,$
Condition 2: $f_i(x_i(t), u_i(t), t) \neq f_i(x_i(t), u_i(t - \tau), t), \; 0 < \tau < \infty$

where x_i is the q_i-vector state, u_i the r_i-vector decision of DM i, and z_i is the k_i-vector information available to DM i and are all a function of time t. The functions f_i and h_i are in general nonlinear and define $f_i : \mathbb{R}^{q_i} \times \mathbb{R}^{r_i} \to \mathbb{R}^{q_i}$ and $h_i : \mathbb{R}^{q_1} \times \mathbb{R}^{q_2} \times \cdots \times \mathbb{R}^{q_N} \to \mathbb{R}^{k_i}$ where $\mathbb{R}^n$ denotes an n-vector of real numbers.

Conditions 1 and 2 ensure that all decisions are nontrivial and each decision maker has a finite-time response.

Definition 2 *A collective system, such that all decision makers participate in a common scalar payoff $J(x, u, t)$, is termed* **collaborative***.*

Collaboration by itself is not a very restrictive condition. A collaborative team need not have common information or rule set. Indeed, the only thing collaborating team members do share is the payoff, the specifics of which they may not even be aware of. Narrowing the scope of collaboration a bit, is complementation.

Definition 3 *Consider the decisions*

$u(t) = [u_1(t), \ldots, u_N(t)]$, $u_i(t) = \gamma_i(z_i(t), t)| \; \gamma_i \in \Gamma_i$. *If for all* $\gamma_i \in \Gamma_i$ *there exists an admissible* $\gamma_j \in \Gamma_j$, $j \neq i$, *then* u_i *is called a complementary decision. If* u_i *are complementary for all* $i = 1, \ldots, N$, *the system of decisions* $u(t)$ *is termed complementary.*

Definition 4 *A collaborative team having a complementary system of decisions is called* **complementary**.

Notice that complementation does not in any way imply communication. While complementary rule sets do not interfere with one another, they may do so in a predetermined or perhaps serendipitous manner. Herein lies the key for distinguishing cooperation. But first we must define what is meant by sensing and communication.

Definition 5 *Consider the collaborative entities defined by*

$$x(t) = [x_1(t) \ldots, x_N(t)],$$
$$u(t) = [u_1(t), \ldots, u_N(t)], \; u_i(t) = \gamma_i(y_i(t), t), \; i = 1, \ldots, N,$$

where y_i *is an* m_i*-vector of measurements. Consider* $h_{ij} : \mathbb{R}^{q_j} \to \mathbb{R}^{m_{ij}}$ *such that* $\sum_j m_{ij} = m_i$.

If $y_i(t) = h_{ij}(x_j(t), t) \neq 0$ *for* $j \neq i$, *then DM* i *is said to sense DM* j. *If* $y_i(t) = [h_{ij}(x_j(t), t) \neq 0 \; \forall j \in \mathbb{H}_i]$ *where* $\mathbb{H}_i \subseteq [1, 2, \ldots, N]$ *and such that* $\mathbb{H}_i \neq \emptyset$, *then DM* i *is said to be sensing. If DM* i *is sensing for all* $i = 1, \ldots, N$, *the system of DMs is termed sensing. Sensing may take one of 2 fundamental forms.*

Case 1. *Autonomous sensing occurs if DM* i *senses DM* j *without any assistance (or perhaps knowledge) from DM* j.

Case 2. *Cooperative sensing occurs when the observed entity, say DM* j, *performs some task that is solely designed to assist the observing entity DM* i *in determining* $h_{ij}(x_j(t), t)$.

In Definition 5, $y_i(t)$ has been substituted for the $z_i(t)$ presented in Definition 1, the reasons to be made clear in Definition 6. A subscript j was added to h, as in h_{ij}, to reflect the measurement coupling between DMs i and j. The subset of DM's coupled into DM i's measurements are in the integer set defined here as $\mathbb{H}_i$. Cooperative sensing is of particular interest since the extreme case, whereupon the observed entity transmits an observation of its own state to the observing entity, is what is defined here as communication.

Definition 6 *Consider the collaborative entities*

$$x(t) = [x_1(t) \ldots, x_N(t)].$$

Suppose $y_j(t) = h_{jj}(x_j(t), t) \neq 0$ and define z_i as the p_i-vector of messages at DM i. Then $z_i(t) = d_{ij}(y_j(t), t) \neq 0$ for $j \neq i$ is a communicated message of the state of DM j received by DM i, where d_{ij} denotes the transmission function from DM i to DM j, is defined similarly to h_{ij} in Definition 5, and may model the pertubations due to the transmitter, channel, and reciever. This idea will now be generalized.

General Definition. *Define the measurements of DM j excluding any sourced at i as*

$$y_{j \backslash i}(t) = y_j(t) : \ \mathbb{H}_j = \{\mathbb{H}_j - i\}.$$

Then $z_i(t) = d_{ij}(y_{j \backslash i}(t), t) \neq 0$ for $j \neq i$ is termed a communicated message of all of the observations of DM j to DM i <u>without redundancy</u>.

If $z_i(t) = [d_{ij}(y_{j \backslash i}(t), t) \neq 0, \ \forall j \in \mathbb{D}_i]$ where $\mathbb{D}_i \subseteq [1, 2, \ldots, i-1, i+1, \ldots N]$ and such that $\mathbb{D}_i \neq \emptyset$, then DM i is said to be communicating. If DM i is communicating for all $i = 1, \ldots, N$, the system of DMs is termed <u>communicating</u>.

Notice that $y_{j \backslash i}(t)$ is defined to prevent needless redundancy in the communications of measurements. $\mathbb{D}_i$ is an integer set that defines the transmission coupling between DM i and all other DMs. It is fairly clear that all communications are due to coupled measurements, implying the standard notion that states must be observed prior to being transmitted.

By the general definition, if every entity is receiving a message, then the system is communicating. As seen in the general definition, communication can be much more extensive in scope than sensing, as observations from DM k as made by DM j may be transmitted from DM j to i. In the definitions of sensing and communication we ignored the trivial case where $h_{ij}(x_j(t), t) = 0$ and $d_{ij}(x_j(t), t) = 0 \ \forall t, j$. In general it may be true that $h_{ij}(x_j(\tau), \tau) = 0$ and $d_{ij}(x_j(\tau)\tau) = 0$ where τ is a point or finite interval on $t \subset \mathbb{R}$ and that τ need not be unique. This suggests the notion of static versus dynamic information patterns.

Definition 7 *If there exists a τ, possibly not unique, such that*

$$h_{ij}(x_j(t), \tau) = 0 \quad (likewise \ d_{ij}(x_j(t), \tau) = 0),$$

then sensing (communication) of DM j at (to) DM i is called **dynamic.**
Otherwise, it is termed **static.**

Essentially, Definition 7 expresses that if H_i (D_i) is time varing, then
sensing (communication) is dynamic. The definition of cooperation now
follows naturally.

Definition 8 *A collaborative system of decision makers that is communicating is called* **cooperative.**

There are two additional classes in the collective systems taxonomy
that merit discussing. The first is based on the case of autonomous sensing. The process of a decision maker autonomously sensing the states of
other decision makers closely resembles that of cooperation. However,
autonomous sensing is a weaker condition in that, as previously discussed, communication permits a decision maker to receive observations
of many entities from just one entity. This ability for a communication
network to pass messages along from one node to another without direct correspondence is known as *multihop*. Autonomous sensing is not
capable of multihop and is therefore weaker. Of course a definition of
autonomous sensing systems analogous to that of cooperation may be
constructed. Since this is not important for the current discussion of
cooperative systems, this work will be deferred for now.

the last class of interest is the overlap of complementary and cooperative systems which are defined as *coordinated* systems. Observe that a
complementary system that is not coordinated is now easily recognized as
having the properties widely associated with *emergent* systems, that is,
a complementary rule set without the benefit of communication. Figure
9.3 depicts the taxonomy of collective systems in a Venn form.

2.1.   The Coupling Equations.

With the model of collective systems firmly in place, the next course
of action is to develop the equations that describe the state evolution of
a collective system. Once accomplished, it should be straightforward to
attribute different classes of the taxonomy to specific components of the
equations. Begin with the state equations presented in Definition 1 and
add the concepts of sensing and communication described in Definitions
5 and 6.

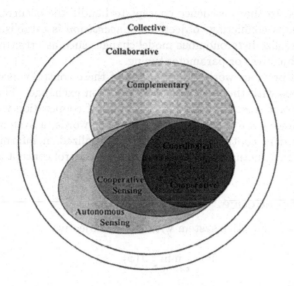

Figure 9.3. Taxonomy of Collective Systems

For $i = 1, \ldots, N$, $N \geq 2$

minimize $J_i(x_i(t), u_i(t), t)$ (14)

$$\dot{x}_i(t) = f_i(x_i(t), u_i(t), t), \tag{14a}$$

$$y_i(t) = [h_{ij}(x_j(t), t) \ \forall j \in \mathbb{H}_i] \text{ where } \mathbb{H}_i \subseteq [1, 2, \ldots, N] \tag{14b}$$

$$y_{j \backslash i}(t) = y_j(t) : \ \mathbb{H}_j \subseteq [1, 2, \ldots, i-1, i+1, \ldots N] \tag{14c}$$

$$z_i(t) = [d_{ij}(y_{j \backslash i}(t), t), \ \forall j \in \mathbb{D}_i, \ u_k, \ \forall k \in \mathbb{C}_i] \tag{14d}$$

$$\text{where } \mathbb{D}_i \subseteq [1, 2, \ldots, i-1, i+1, \ldots N]$$

$$u_i(t) = \gamma_i(y_i(t), z_i(t), t) : \ \gamma_i \in \Gamma_i \neq \emptyset \tag{14e}$$

where u_k, $\forall k \in \mathbb{C}_i$ was added to $z_i(t)$ to ensure equation (14d) generalizes (7).

3. Precedence, hierarchy, and supervision

Three control concepts are now introduced, each having some history in team theory, organizational theory, and decentralized control, but now united within the single framework of collective systems. Each of these concepts relies to some degree on the transfer of information.

Precedence will continue to be defined as first introduced in team theory but will be clarified somewhat in notation. *Hierarchy* is introduced as an extension to the precedence concept to handle the occurrence of multiple, perhaps conflicting, objectives. *Supervision* is established as the means to realize heteronomous model changes such as artificial constraints and objectives or parameter tuning.

Unlike most previous work, the structure of these control concepts is not synonymous with the structure of information exchange. This is a fundamental point since it raises the possibility of cooperation without any form of precedence or supervision. In other words, a system may be decentralized in control but completely centralized in information. The information structure will be related to each control concept as it is addressed.

3.1.   Precedence.

Consider a collaborative system with shared objective

$$\textbf{P1} \quad \min_{\text{s.t. } \phi \in \Phi} \ J(\phi)$$

where ϕ is shorthand notation for the state variables of (14a) - (14e) and Φ is the admissible set.

First assume that each decision maker solves **P1** without explicit knowledge of another's decision. Observe that this does not mean that they cannot exchange information, either through autonomous sensing or communication (i.e. $\mathbb{H}_i \neq \emptyset$, $\mathbb{D}_i \neq \emptyset$ in Definitions 5 and 6). As previously understood, the precedence relation for this condition would have an information structure precedence diagram with N nodes and no edges despite the fact that information may be transfered. Clearly a new relational graph structure is warranted to help relate and contrast precedence and information. The example in Figure 9.4 shows a system with no precedence and an information coupling defined by

$$d_{ij} \neq 0, \ \forall j \in \mathbb{D}_i \text{ where } D_2 = [1,3], D_3 = [1,2], D_4 = [5], D_5 = [4].$$

The information exchange could be any digraph; the one depicted is for illustrative purposes only. The figure is a directed *multi graph* in that there are two sets of directed edges on a common set of vertices [7]. The subgraph on the left is defined as the *precedence structure* and the subgraph on the right as the *communication structure*. Combined they form a complete picture of the coupling for this cooperative system. Of course additional subgraphs may be added to represent autonomous sensing, hierarchy, and supervision. The resulting multi graph will be termed the *coupling structure*. Before proceeding to fully define the coupling structure, the precedence structure will be clarified.

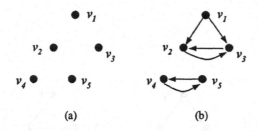

(a)   (b)

Figure 9.4. Example of a Coupling Structure Multi Graph With No Precedence: (a) Precedence Structure; (b) Communications Structure

Returning to the example, assume that each decision maker solves **P1** with precedence as dictated by Figure 9.5a. Then the effect is for DM 1 to solve **P1** for u_1^* and pass it along to DMs 2 and 3. They in turn solve **P1** given u_1^* after which DM 2 passes u_2^* on to DMs 4 and 5. Once again **P1** is solved, this time by DMs 4 and 5 given u_2^* only.

If DM 2 were to pass along u_1^* with u_2^*, DM 2 would have *memory* of prior decisions (multihop) which is indicated in Figure 9.6a by the dotted lines on (v_2, v_4) and (v_2, v_5). Unlike the Ho and Chu model, the dotted line does not necessarily imply nesting of z_j in z_i. For it may be true that $j \in \mathbb{C}_i$ while $j \notin \mathbb{D}_i$ in equation (14d). However, since in general $\mathbb{C}_i \neq \mathbb{D}_i$, z_j need not be nested in z_i to guarantee the partially nested condition. For the nesting condition to be true, only those components of z_j that belong to the precedents of DM i need be nested in z_i. The stronger nesting condition of z_j in z_i may have greater implications and will be studied in further work.

Instead of relying on the memory of DM 2 for delivering u_1^* to DMs 3 and 4, DM 1 may pass u_1^* to them directly as indicated in Figure 9.7a.

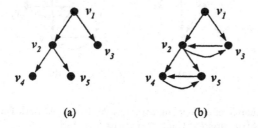

(a)   (b)

Figure 9.5. Example of a Coupling Structure Multi Graph With Precedence: (a) Precedence Structure; (b) Communication Structure

Now consider what the precedence structure implies on the communications structure and is borne out in Figures 9.5b, 9.6b. and 9.7b.

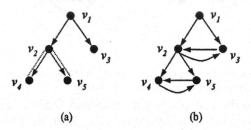

Figure 9.6. Example of a Coupling Structure Multi Graph With Memory: (a) Precedence Structure; (b) Communications Structure

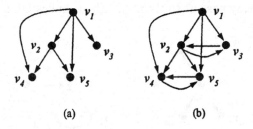

Figure 9.7. Example of a Coupling Structure Multi Graph With Direct Precedence: (a) Precedence Structure; (b) Communications Structure

Proposition 1 *The precedence structure graph is a spanning subgraph of the communications structure graph.* **Proof** *The spanning condition follows from the definition of a multi graph. The relation j precedes i implies a directed edge from j to i on the precedence subgraph. Since u_j^* are contained in $z_i(t)$ per equation (14d), a directed edge from j to i on the communications subgraph is necessary to ensure each relationship of j precedes i.* ∎

The proof provides only a necessary condition on the communications subgraph. The sufficient condition is for $j \in \mathbb{C}_i$. This need not be the case in general, hence the bidirectional arcs on (v_2, v_3) and (v_4, v_5).

3.2. Multiple Objectives, Caucuses and Hierarchy.

In equation (14) allowances are made for each individual decision maker to have dissimilar, perhaps conflicting, objectives. Individual conflict may be the result of a common, collaborative objective being at odds with individual objectives. Conflict may also arise when the decision maker is a member of more than one collaborative *caucus*. It can be argued that the first situation is nothing more than a special case of the second with the individual objective equivalent to a caucus with membership of one. Nevertheless, when there exists more than one caucus of objectives, collaboration assumes a complex form. If the underlying objectives are prioritized, a *caucus hierarchy* results.

Consider the collaborative system attempting to solve **P1** that was introduced in the previous section example to be a single caucus. Suppose two more caucuses of decision makers were added such that the 3 caucus objectives are now

$$\text{Caucus 1. } \mathbf{P1} \quad \min_{\text{s.t. } \phi_1 \in \Phi_1} J(\phi_1)$$

$$\text{Caucus 2. } \mathbf{P2} \quad \min_{\text{s.t. } \phi_2 \in \Phi_2} J(\phi_2)$$

$$\text{Caucus 3. } \mathbf{P3} \quad \min_{\text{s.t. } \phi_3 \in \Phi_3} J(\phi_3)$$

A trivial case occurs when $\Phi_1 \cap \Phi_2 = \emptyset$, $\Phi_1 \cap \Phi_3 = \emptyset$ and $\Phi_2 \cap \Phi_3 = \emptyset$. In words, none of the decision makers have more than one objective. In this case, the three collaborative systems are independent and may be treated individually. In general, if $\Phi_i \cap \Phi_j = \emptyset$ for $j = 1, \ldots, V$, where V is the total number of objectives in the system, then caucus i may be treated independently.

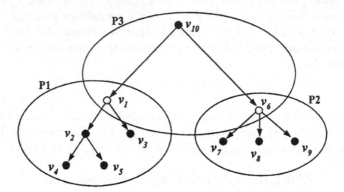

Figure 9.8. Example of Flat Network of Caucuses

Now consider the case where $\Phi_1 \cap \Phi_2 = \emptyset$, $\Phi_1 \cap \Phi_3 \neq \emptyset$ and $\Phi_2 \cap \Phi_3 \neq \emptyset$, and $\Phi_1 \cap \Phi_2 \cap \Phi_3 = \emptyset$. An example is shown in Figure 9.8. In the figure, there are 10 nodes with edges representing precedence per the definition previously provided. The ellipses in the diagram each uniquely correspond to one caucus as labeled. The members of each caucus are enclosed by the corresponding ellipse. Hence the caucuses for this example are $(v_1, v_2, v_3, v_4, v_5)$, (v_6, v_7, v_8, v_9), (v_1, v_6, v_{10}). The figure illustrates what might be an example of a caucus of decision makers participating in a high level decision process **P3** while two other caucuses work on lower level problems **P1** and **P2**. For this example, nodes v_1 and v_6 represent mid-level managers.

The mid-level managers must deal with possible conflict, as indicated by their being enclosed by two ellipses and being shaded white in the diagram. Henceforth we will call vertices with membership in multiple caucuses *conflict nodes*. At issue is how to represent the conflict, both graphically and mathematically. Taking a cue from multi criteria optimization, there are two ways one might solve the problems presented to decision makers 1 and 6. The first is to solve a weighted sum of the

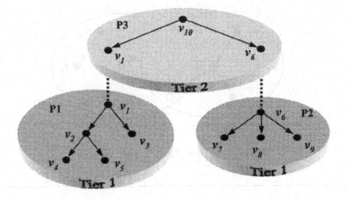

Figure 9.9. Example of a Hierarchical Network of Caucuses

objective as in

$$\textbf{M1} \qquad \min \quad \lambda J(\phi_1) + (1 - \lambda)J(\phi_3)$$
$$\text{s.t. } \phi_1 \in \Phi_1$$
$$\text{and } \phi_3 \in \Phi_3$$

for v_1 and

$$\textbf{M2} \qquad \min \quad \lambda J(\phi_2) + (1 - \lambda)J(\phi_3)$$
$$\text{s.t. } \phi_2 \in \Phi_2$$
$$\text{and } \phi_3 \in \Phi_3$$

for v_6.

with $0 \leq \lambda \leq 1$. This approach will result in what is denoted a *flat network*.

The alternative is to rank the objectives for the conflict nodes so that they are able to solve them in order of priority. This approach leads to a *caucus hierarchy*. Our example is shown in this form in Figure 9.9, where decision maker 1 and 6 must each first solve **P3** before proceeding on to **P1** and **P2** respectively. In the figure, priority is indicated graphically via the use of the disk shaped *tiers*. The higher the tier in the graph (both numerically and visually), the greater the priority. Hence **P3** is a higher priority than either **P1** or **P2** which are equal in priority as far as we are concerned. If $\Phi_1 \cap \Phi_2 \neq \emptyset$, then there may be a need for determining a priority ranking between **P1** and **P2**. In both diagrams, the representation of information flow remains unchanged from previously described.

Of course multi criteria optimization is nothing new. What makes these problems unique is the additional complication of precedence. So,

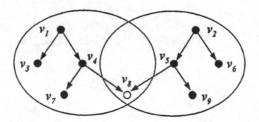

Figure 9.10. The "Two Bosses" Example

while the situation in Figures 9.8 and 9.9 appear reasonable enough, many other possibilities exists, such as the example in Figure 9.10 where decision maker 3 has two bosses. For, while solving multi criteria problems in general, introduces the risk that an admissible solution does not exist, adding precedence, *especially when the conflict nodes are at the bottom of a precedence relation* as in Figure 9.10, makes matters much worse.

3.3.   Supervision

At the end of Section 1.1 an idea was presented that might make the design of systems with non nested information structures easier to consider. The idea was to introduce artificial objectives, constraints, or parameter values to a subordinate to overcome his lack of information. This concept, termed *heteronomous modeling* , will now be formalized through the use of a *supervision structure* subgraph on the coupling multi graph.

Returning to the example of Section 3.1, consider the system

$$\textbf{P1} \quad \min_{\substack{\text{s.t. } \phi \in \Phi}} J(\phi)$$

with precedence as defined in Figure 9.5a. Now suppose that instead of adding memory to the system as suggested by Figure 9.6a or sending decisions directly as in Figure 9.7a, DM 1 provides an alternate *artificial problem* to DMs 2 and 3 and DM 2 follows suit with DMs 4 and 5. Let the artificial problems be defined as

$$\text{Master problem. } \mathbf{P1} \quad \min_{\phi \in \Phi} \ J(\phi)$$

$$\text{DM 2 artificial problem. } \mathbf{P1}^{a_2} \quad \min_{\phi \in \Phi^{a_2}} \ J^{a_2}(\phi)$$

$$\text{DM 3 artificial problem. } \mathbf{P1}^{a_3} \quad \min_{\phi \in \Phi^{a_3}} \ J^{a_3}(\phi)$$

$$\text{DM 4 artificial problem. } \mathbf{P1}^{a_4} \quad \min_{\phi \in \Phi^{a_4}} \ J^{a_4}(\phi)$$

$$\text{DM 5 artificial problem. } \mathbf{P1}^{a_5} \quad \min_{\phi \in \Phi^{a_5}} \ J^{a_5}(\phi)$$

If the artificial problems guarantee the optimal solution to the original problem **P1**, then the new system has effectively used supervision and is called *supervised optimal*. The nice property of supervision is that it offers us the ability to overcome the limitations induced by non partially nested information structures and holds the potential to extend us beyond the linear models and quadratic performance indices therein. Even if artificial problems that guarantee the optimal solution cannot be developed, they can be designed with bounds on the optimal solution via relaxations. The same cannot be said for systems with only precedence. As an example, the artificial problems may have a discrete, finitely countable support. If so, the system is readily modeled as a *hybrid discrete event system* and is now amenable to the many tools at our disposal for solving these types of problems [2, 3, 5, 10, 21, 22]. This is a topic of further research and will not be covered in this overview.

Supervised systems with multiple objectives admit hierarchies just as systems with precedence do. The communication structure induced by the supervised hierarchy is also developed in the same manner as that for precedence systems. These topics will not be covered since the approach is analogous to that provided in Section 3.2.

Precedence and supervision are two different approaches for influencing a collaborative decision process. They may both be present at the same time within a system. They may complement one another or work against each other. The design of systems with precedence and supervision must be undertaken with great care. General rules for proceeding is a topic for further research.

4.   Summary

This introduction to cooperative systems was mainly motivated by team theory and the deficiency of that theory which requires the information of one decision maker to be nested within that of his precedent to guarantee an affine solution form. If not affine, there are no formal methods for constructing the non-affine form, hence we cannot even construct reasonable bounds. The nature of cooperation was revealed by introducing it within the framework of a broader class of problems, namely, the Collective systems. Along the way, models for many simpler systems were developed that clearly put into perspective the difference between true cooperation and the appearance of cooperation. Based upon this general framework, concepts of precedence were expanded from team theory and the role of communication broadened. This allowed the presentation of collective systems with multiple objectives as a collection of caucuses, each vying for resources. Supervision was introduced as an alternative to precedence that uses the manipulation of a subordinate's model or parameters to ensure optimality without the presence nested information.

There are many topics that must be further explored before this line of research begins to bear fruit. The nature of complete communications nesting z_j in z_i for collaborative systems must be explored. Complete nesting is a much broader condition than nesting for precedence only, as considered by Ho and Chu [8]. Besides cooperation, other classes of the collective systems taxonomy must be examined in greater depth, especially complementary, coordinated, and autonomous sensing systems. The design of supervisory artificial objectives, constraints and switching surfaces should be examined which likely implies that an understanding of the interaction of precedence and supervision is needed.

Acknowledgments

The author wishes to thank the many participants at the December 2000 Workshop For Cooperative Control and Optimization, Gainesville, FL for their many ideas and comments during the open discussion session. This chapter is the product of that discourse and would not have been possible without the honest and forthright comments of the author's colleagues.

References

[1] Basar T. and Olsder G. (1982). *Dynamic Noncooperative Game Theory*. Academic Press, NY.

[2] Branicky M. (1998). "Multiple Lyapunov functions and other analysis tools for switched and hybrid systems, " IEEE Transactions on Automatic Control, Vol. 43, No. 4, Apr 1998, 475 - 482.

[3] Branicky M., Borkar V., and S. Mitter (1998). "A unified framework for hybrid control: model and optimal control theory," IEEE Transactions on Automatic Control, Vol. 43, No. 1, Jan 1998, 31 - 45.

[4] Deng X. and Papadimitriou C. (1999). "Decision making by hierarchies of discordant agents," Mathematical Programming, Vol. 86, No. 2, 417-431.

[5] Di Febbraro A. and Sacone S. (1998). "Hybrid petri nets for the performance analysis of transportation systems," Proc. 37th IEEE Conference on Decision and Control, Tampa, FL.

[6] Geanakoplos J.and Milgrom P. (1991)."A theory of hierarchies based on limited managerial attention," J. of the Japanese and International Economies, Vol. 5, No. 3, 205-225.

[7] Harary F. (1969). "Graph Theory", Addison-Wesley Publishing Company, Reading, MA.

[8] Ho Y.C. and Chu K.C. (1972). "Team decision theory and information structures in optimal control problems-part 1," IEEE Trans. Automatic Control, Vol. AC-17, No. 1, 15-20.

[9] Ho Y.C. and Chu K.C. (1974). "Information structure in dynamic multi-person control problems," Automatica, Vol. 10, 341-351.

[10] Hubbard P. and Caines P. (1999). "Initial investigations of hierarchical supervisory control for multi-agent systems," Proceedings of

the 38th Conference on Decision and Control, Pheonix, AZ USA, Dec 1999, 2218 - 2223.

[11] Kuhn R. (1997). "Sources of failure in the public switched telephone network," IEEE Computer, April 1997.

[12] Locsin J. (1994). "SCADA/EMS study reflects changing industry," Electric Light and Power, Sep. 1994, 35-37.

[13] Marschak J. (1955). "Elements for a theory of teams," Management Science, Vol. 1, 127-137.

[14] (1972). Marschak J. and Radner R. *Economic Theory of Teams.* Yale University Press, New Haven CT.

[15] Marschak T. and Reichelstein S. (1998). "Network mechanisms, informational efficiency, and hierarchies," J. of Economic Theory, 79, 106-141.

[16] McDonald J. (1994). "Public network integrity–avoiding a crisis in trust," IEEE Journal on Selected Areas in Communications, Jan 1994, 5-12.

[17] North American Electric Reliability Council (NERC) (1985). "Reliability Concepts," Princeton NJ.

[18] North American Electric Reliability Council (NERC) (1996). "Reliability Assessment 1996-2005: The Reliability of Bulk Electric Systems in North America," Princeton NJ.

[19] Ownbey P., Schaumburg F. and Klingeman P. (1988). "Ensuring the security of public water supplies," Journal of the American Water Works Association, Feb. 1988, 30-34.

[20] Radner R. (1962). "Team decision problems," Ann. Math Statistics, Vol 33, No. 3, 857-881.

[21] Ramadge P. and Wonham W. (1989). "The control of discrete event systems," Proceedings of the IEEE, Vol. 77, No. 1, Jan 1989, 81 - 98.

[22] Rudie K. and Wonham W. (1992). "Think globally, act locally: decentralized supervisory control," IEEE Transactions on Automatic Control, Vol. 37, No. 11, Nov 1992, 1692 - 1708.

[23] Smith J. (1995). "Disaster avoidance and recovery," Government Computer News, Jun. 19, 1995, 67-69.

[24] U.S. Federal Highway Administration, (1995). "Status of the Nation's Surface Transportation System: Condition and Performance," Report to Congress, 1996.

[25] U.S. General Accounting Office, (1993). "Telecommunications: Interruptions of Telephone Service," Report GAOO/RCED-93-79FS, Mar. 1993.

[26] Witsenhausen H.S. (1968). "A counterexample in stochastic optimum control," Siam J. Control, Vol. 6, No. 1, 131-147.

Chapter 10

COOPERATIVE AIRCRAFT CONTROL FOR MINIMUM RADAR EXPOSURE *

Meir Pachter *
Air Force Institute of Technology, Wright-Patterson AFB OH, USA
meir.pachter@afit.edu

Jeffrey Hebert †
jeffrey.hebert@afit.edu

Abstract   Two aircraft exposed to illumination by a tracking radar are considered
and the optimization problem of cooperatively steering them to a prespecified rendezvous point is addressed. First, the problem of a single
aircraft exposed to illumination by a tracking radar is considered and
the problem of determining an optimal planar trajectory connecting two
prespecified points is addressed. The solution is shown to exist only if
the angle θ_f, formed by the lines connecting the radar to the two prespecified trajectory points, is less than 60°. In addition, expressions
are given for the optimal path length, l^*, and optimal cost. When the angle
$\theta_f \geq 60°$, an unconstrained optimal solution does not exist, and in order
to render the optimization problem well posed, a path length constraint
is imposed. Numerical optimization techniques are used to obtain optimal aircraft trajectories for the constrained case. Finally, the problem
of isochronous rendezvous of the two aircraft is addressed using an optimization argument and the analytic results previously derived for a
single aircraft trajectory.

Keywords: cooperative control, radar exposure

---

*The views expressed in this article are those of the authors and do not reflect the official
policy of the U.S. Air Force, Department of Defense, or the U.S. Government.
*Professor, Dept. of Electrical and Computer Engineering
†Captain, U.S. Air Force

R. Murphey and P.M. Pardalos (eds.), Cooperative Control and Optimization, 199–211.
© 2002 *Kluwer Academic Publishers.*

1.   Single vehicle radar exposure minimization

Given a radar located at the origin O of the Euclidean plane, it is desired to find the optimal aircraft trajectory that connects two prespecified points A and B in the plane such that the Radio Frequency (RF) energy reflected from the aircraft is minimized; see, e.g., Fig. 10.1. According to the "Radar Transmission Equation"[4], the ratio of the received RF power to the transmitted RF power reflected from the target is inversely proportional to R^4, where R is the slant range from the target to the monostatic radar. The cost to be minimized is then

$$\int_0^{\frac{l}{v}} \frac{1}{R^4(t)} dt$$

where v is the (constant) speed of the aircraft and l is the path length. Now, consider the trajectory in Fig. 10.1 to be given in polar form, as $R = R(\theta)$. Furthermore, $v = \frac{ds}{dt}$ i.e., $dt = \frac{ds}{v}$, and ds, the element of arc length, is given in polar coordinates by

$$ds = \sqrt{\left(\frac{dR}{d\theta}\right)^2 + R^2} \, d\theta$$

Substituting into the cost equation we then obtain the functional

$$J[R(\theta)] = \int_0^{\theta_f} \frac{\sqrt{\dot{R}^2 + R^2}}{R^4} \, d\theta \tag{1}$$

The boundary conditions are

$$R(0) = R_o \tag{2}$$

$$R(\theta_f) = R_f, \quad 0 < \theta \le \theta_f. \tag{3}$$

Without loss of generality, assume $R_f \ge R_o$ and $0 < \theta_f \le \pi$, see, e.g., Fig. 10.1. Polar coordinates are used. We have the following:

Theorem 1 *The optimal trajectory which connects points A and B at a distance R_o and R_f from the radar located at the origin O, where θ_f is the angle $\angle AOB$, and minimizes the exposure to the radar according to Eqs. (1)-(3), is*

$$R^*(\theta) = \frac{R_o}{\sqrt[3]{\sin \phi}} \sqrt[3]{\sin(3\theta + \phi)}, \ 0 < \theta \le \theta_f \tag{4}$$

where the angle

$$\phi = \arctan \left(\frac{\sin 3\theta_f}{\left(\frac{R_f}{R_o}\right)^3 - \cos 3\theta_f} \right)$$

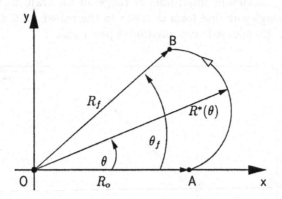

Figure 10.1. Optimal Trajectory

Moreover, the length of the optimal path is given by the integral

$$l^* = \frac{R_o}{\sqrt[3]{\sin\phi}} \int_0^{\theta_f} [\sin(3\theta + \phi)]^{-\frac{2}{3}} \, d\theta \tag{5}$$

and the cost function explicitly evaluates to

$$J^* = \frac{1}{3R_o{}^3} \frac{\sin 3\theta_f}{\sin(3\theta_f + \phi)}$$

This result holds provided $0 < \theta_f < \frac{\pi}{3}$. However, if $\frac{\pi}{3} \leq \theta_f \leq \pi$, then an optimal path does not exist and a constraint on the path length, l, must be included to render the optimization problem well posed.

Proof. See [3].

The optimal trajectory given in Eq. (4) is shown in Fig. 10.2 for the case where $R_o = R_f = 1$ and $\theta_f = 45°$. This trajectory is related to a class of functions known as rose functions or *Rhodenea* [1]. In polar coordinates, a three-leaved rose function can be written as

$$R = a\sin(3\theta + \phi)$$

Since $0 < \theta_f < \pi/3$ for our problem, evidently we are concerned only with the "leaf" in the first quadrant. The parameter ϕ rotates the three-leaved rose function about the origin and the constant

$$a = \frac{R_o}{\sqrt[3]{\sin\phi}}$$

describes the maximum amplitude, or range, of the trajectory from the origin. Although our sine term is taken to the cubed root, the general shape of the three-leaved rose function is preserved.

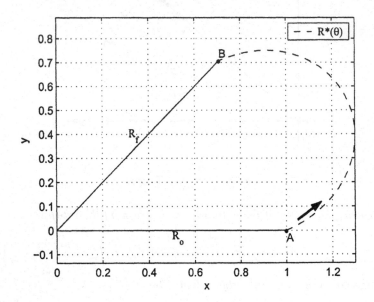

Figure 10.2. Optimal Trajectory $R^*(\theta)$ for $R_o = R_f = 1$ and $\theta_f = 45°$

Remark 1 *Equivalently, the optimal trajectory can be written as*

$$R^*(\theta) = \sqrt[3]{\frac{R_f^3 - R_o^3 \cos 3\theta_f}{\sin 3\theta_f} \sin 3\theta + R_o^3 \cos 3\theta}$$

and the optimal cost can be written as

$$J^* = \frac{\sqrt{R_f^6 + R_o^6 - 2R_o^3 R_f^3 \cos 3\theta_f}}{3R_o^3 R_f^3}$$

where ϕ has been eliminated.

The expression for the optimal path length, l^*, Eq. (5), is not given in closed form. To evaluate this integral, numerical integration can be employed. Additionally, this integral can be calculated using tables of elliptic integral functions found in standard mathematics reference texts

[2] or as function calls in mathematics software packages [5]. We have the following:

Proposition 1 *The path length, l^*, of the optimal trajectory minimizing the exposure to a monostatic radar given in Theorem 1, formulated using the Legendre form of the elliptic integrals, is*

$$l^* = \begin{cases} \frac{R_o}{2\sqrt[4]{3}} \frac{1}{\sqrt[3]{\sin\phi}} \left[4K(k) - F(\psi_1, k) - F(\psi_2, k)\right], & 3\theta_f + \phi > \frac{\pi}{2} \\ \frac{R_o}{2\sqrt[4]{3}} \frac{1}{\sqrt[3]{\sin\phi}} \left[F(\psi_1, k) - F(\psi_2, k)\right] & , & 3\theta_f + \phi \leq \frac{\pi}{2} \end{cases}$$

where $F(\psi_i, k)$ are incomplete elliptic integrals of the first kind, $K(k)$ is a complete elliptic integral of the first kind and

$$k = \frac{\sqrt{2 - \sqrt{3}}}{2}$$

$$\psi_1 = \arccos\left(\frac{1 - (1 + \sqrt{3})\left[\sin(3\theta_f + \phi)\right]^{2/3}}{1 + (\sqrt{3} - 1)\left[\sin(3\theta_f + \phi)\right]^{2/3}}\right)$$

$$\psi_2 = \arccos\left(\frac{1 - (1 + \sqrt{3})\left[\sin\phi\right]^{2/3}}{1 + (\sqrt{3} - 1)\left[\sin\phi\right]^{2/3}}\right)$$

Fig. 10.3 demonstrates how the length l^* of the optimal aircraft trajectory varies with the problem parameters θ_f and R_f/R_o. In Fig. 10.3, $l^*(\theta_f, R_f/R_o)$ is plotted for $R_f/R_o = 1, 2, 3, 4$ and $\theta_f = [0, 59°]$ and the asymptotic behavior of the path length, viz., $l^* \to \infty$ as $\theta_f \to 60°$, is evident.

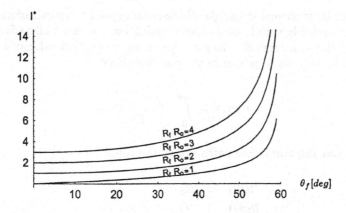

Figure 10.3. $l^*\left(\theta_f, \frac{R_f}{R_o}\right)$ for the Cases where $\frac{R_f}{R_o} = 1, 2, 3, 4$

Although the path length l^* approaches infinity as the angle θ_f approaches 60 degrees, the optimal cost approaches the finite value given by

$$\lim_{\theta_f \to \frac{\pi}{3}} J^* = \frac{1}{3}\left(\frac{1}{R_o{}^3} + \frac{1}{R_f{}^3}\right)$$

as depicted in Fig. 10.4.

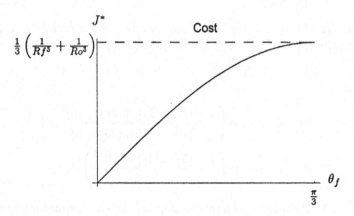

Figure 10.4.   Asymptotic Behavior of the Cost Function J^*

1.1.   Variable path length solutions

In order to extend the single vehicle radar exposure minimization problem to multiple vehicle isochronous rendezvous, we must allow for optimal paths of prespecified length. Assuming a constant velocity for the vehicles, we wish to minimize the cost functional

$$J[R(\theta)] = \int_0^{\theta_f} \frac{\sqrt{\dot{R}^2 + R^2}}{R^4}\, d\theta$$

such that the admissible paths satisfy the conditions

$$
\begin{aligned}
R(0) &= R_o \\
R(\theta_f) &= R_f, \quad 0 < \theta \le \theta_f \\
L[R(\theta)] &= \int_0^{\theta_f} \sqrt{\dot{R}^2 + R^2}\, d\theta = l
\end{aligned}
$$

where $L[R(\theta)]$ is a functional representing the path length constraint. We can express the augmented cost functional as

$$J = \int_0^{\theta_f} \left(\frac{\sqrt{\dot{R}^2 + R^2}}{R^4} + \lambda \sqrt{\dot{R}^2 + R^2} \right) d\theta \qquad (6)$$

where λ is the familiar Lagrange multiplier. The resulting Euler equation yields the nonlinear, ordinary differential equation

$$R''(\theta)[R(\theta) + \lambda R^5(\theta)] + 3R^2(\theta) - \lambda R^6(\theta) + 2[R'(\theta)]^2 - 2\lambda R^4(\theta)[R'(\theta)]^2 = 0$$

which can be expressed as the following system of first order equations

$$Y_1'(\theta) = Y_2(\theta) \qquad (7)$$
$$Y_2'(\theta) = \frac{3Y_1^2(\theta) - \lambda Y_1^6(\theta) + 2Y_2^2(\theta) - 2\lambda Y_1^4(\theta)Y_2^2(\theta)}{Y_1(\theta) + \lambda Y_1^5(\theta)} \qquad (8)$$

with boundary conditions

$$Y_1(0) = R_o \qquad (9)$$
$$Y_1(\theta_f) = R_f \qquad (10)$$

and with λ a real valued constant. We note that when $\lambda = 0$ we have the original unconstrained problem formulation. Although a closed form solution of the Euler equation of the path length constrained radar problem is out of reach, numerical methods are available to study this two point boundary value problem. The shooting method is the most successful at solving this class of problems.

A numerical ODE solver, e.g., employed by the shooting method, provides as its output a finite number of points along the curve $R^*(\theta)$. In order to evaluate the cost functional, J, given in Eq. (6), the following piecewise linear approximation is made.

The optimal trajectory from points A to B is approximated with a series of N straight line segments - see, e.g., Fig. 10.5. Each of these N segments can be expressed as in the two-point form of a line in polar coordinates, and can be written as

$$R(\theta) = \frac{r_1 r_2 \sin(\theta_2 - \theta_1)}{r_1 \sin(\theta - \theta_1) - r_2 \sin(\theta - \theta_2)} \qquad (11)$$

which has the first derivative

$$\dot{R}(\theta) = \frac{r_1 r_2 \sin(\theta_1 - \theta_2)[r_1 \cos(\theta - \theta_1) - r_2 \cos(\theta - \theta_2)]}{[r_1 \sin(\theta - \theta_1) - r_2 \sin(\theta - \theta_2)]^2} \qquad (12)$$

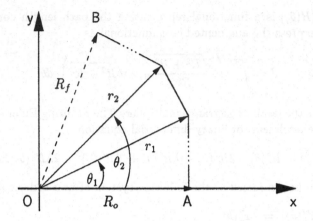

Figure 10.5.   Line Segment Defined by Points (r_1, θ_1), (r_2, θ_2)

For example, the cost to travel from the point (r_1, θ_1) to the point (r_2, θ_2) can be written as

$$
\begin{aligned}
J_{1,2} &= \int_{\theta_1}^{\theta_2} \left(\frac{\sqrt{R^2 + \dot{R}^2}}{R^4} + \lambda \sqrt{R^2 + \dot{R}^2} \right) \, d\theta \\
&= \int_{\theta_1}^{\theta_2} \frac{\sqrt{R^2 + \dot{R}^2}}{R^4} \, d\theta + \lambda \int_{\theta_1}^{\theta_2} \sqrt{R^2 + \dot{R}^2} \, d\theta
\end{aligned}
\tag{13}
$$

Substituting Eqs. (11) and (12) into the cost function, Eq. (13), yields

$$
\begin{aligned}
J_{1,2} = {} & \frac{\sqrt{r_1^2 + r_2^2 - 2r_1 r_2 \cos \Delta\theta}}{4 \left(r_1 r_2 \sin \Delta\theta \right)^3} \left[4 r_1 r_2 \Delta\theta \left(\cos \Delta\theta - \sin \Delta\theta \right) \cdots \right. \\
& \left. - \left(r_1^2 + r_2^2 \right) \left(2\Delta\theta - \sin 2\Delta\theta \right) \right] - \lambda \sqrt{r_1^2 + r_2^2 - 2 r_1 r_2 \cos \Delta\theta}
\end{aligned}
$$

where

$$
\Delta\theta = \theta_1 - \theta_2
$$

Thus we have eliminated any dependence upon θ and the cost of any given line segment is explicitly determined for a given pair of points (r_1, θ_1), (r_2, θ_2). Since the total cost of the optimal path for N line segments is the summation of the costs for each segment, viz.,

$$
\tilde{J}^* = \sum_{i=1}^{N} J_{i,i+1}
$$

One approximation strategy is to increase the number of points, N, until a user specified termination criterion is met, e.g., increase N, until

$$\left| \sum_{i=1}^{N} J_{i,i+1} - \sum_{i=1}^{N-1} J_{i,i+1} \right| < \epsilon$$

where ϵ is the user defined tolerance on the accuracy of $\tilde{J}^*$. A similar strategy can be used to approximate the path length, where

$$\tilde{l}^* = \sum_{i=1}^{N-1} \sqrt{r_1{}^2 + r_2{}^2 - 2r_1 r_2 \cos(\theta_i - \theta_{i+1})}$$

The piecewise linear approximation accurately portrays the approximation to the cost function that is taking place. The variational problem can now be approximated using discrete straight line segments. This permits the resulting problem to be solved using optimization, viz., nonlinear programming algorithms, such as Sequential Quadratic Programming. This approach also facilitates the development of a solution using the finite element method. For the numerical work in this paper, however, the shooting method is used.

2.   Multiple vehicle isochronous rendezvous

Consider the problem of coordinating the isochronous rendezvous of two or more vehicles such that they reach the same point, at the same time, while minimizing their cumulative exposure to a radar.

For each air vehicle, the problem is formulated as a constrained path length problem in the Calculus of Variations. The costs of various path lengths are separately calculated for each vehicle, from a minimum length path, i.e., a straight line, to some maximum length path. Tables of cost vs. time (path length) are then provided by each vehicle to a central decision maker who determines what time of arrival minimizes a composite cost function. The latter is the cumulative exposure to radar of the (two) aircraft. The air vehicles are each provided with an optimal joint time of arrival by the central decision maker, and select the optimal path to perform the rendezvous.

For example, consider a ship who is initially at $(r_1, \theta_1) = (1, 0°)$. Consider a second ship at $(r_2, \theta_2) = (1.5, 10°)$. With a radar located at the origin, the ships must rendezvous at the point $(r_f, \theta_f) = (1, 45°)$. See, e.g., Fig 10.6. For this problem, we'll assume the vehicles move at the same constant speed, thus time of arrival can be equated to path length.

Determining a series of optimal paths of varying length can be accomplished numerically, in this case by employing the shooting method, with Eqs. (7)-(10).

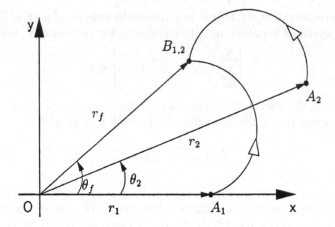

Figure 10.6. Cooperative Isochronous Rendezvous

Remark 2 *The problem of the second ship can be solved by transforming to the equivalent problem, $R_o = 1$, $R_f = 1.5\cos 10°$ and $\theta_f = 35°$. The path would then be traversed backwards in time and shifted by 45°, i.e.,*

$$\theta_i = 45° - \theta_j, \quad i = 1, 2, ..., N, \ j = N, N - 1, ..., 1$$

In Fig. 10.7 we see a curve representing optimal path length versus cost for both vehicles. We note that the optimal path length for vehicle 1 is 1.3229. The optimal path length for vehicle 2 is 1.1646. Fig. 10.8 reflects the cumulative exposure cost function $J_1^* + J_2^*$. Each vehicle is commanded by the central decision maker to plan an optimal path of optimal length 1.2556 and the resulting trajectories are shown in Fig. 10.9. The optimal path length $l^* = 1.2556$ is determined by the central decision maker to minimize the cumulative exposure $J_1^* + J_2^*$.

Proposition 2 *When $\theta_f < 60°$, the optimal unconstrained trajectory exists and curve of cost versus optimal constrained path length is convex. When $\theta_f \geq 60°$, an optimal unconstrained trajectory does not exist and the curve of cost versus optimal constrained path length is not convex.*

3.   Conclusion

The problem of determining the optimal planar flight path such that the exposure of an aircraft to illumination by a tracking radar is minimized, has been solved. Using the insights gained from the single vehicle problem, a hierarchical optimization problem is formulated and the results are extended to determine the optimal trajectories than minimize

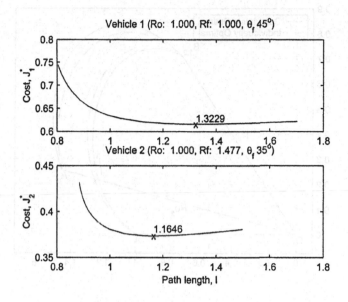

Figure 10.7. Curves of Optimal Cost vs. Path Length for Vehicles 1 and 2

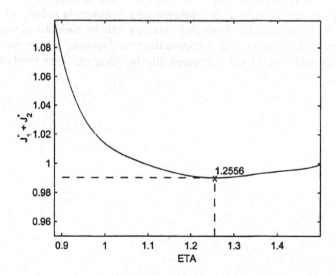

Figure 10.8. Composite Cost Curve Identifying the Optimal Time of Arrival

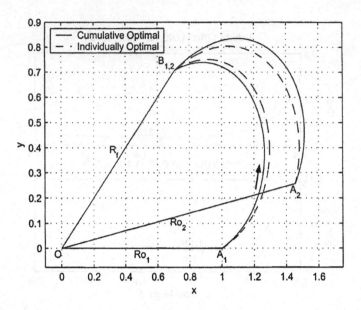

Figure 10.9. Trajectories for the Coordinated Isochronous Rendezvous Example

the cumulative exposure for two vehicles. This, a cooperative control strategy, is implemented. A representative problem is solved which illustrates the approach. A similar strategy can be used to extend this problem to the case of n-ship cooperative rendezvous. Future work will introduce additional levels of complexity building on these fundamental results.

References

[1] Eves, H. (1991). Analytic geometry. In Beyer, W. H., editor, *CRC Standard Mathematical Tables and Formulae*. CRC Press, Boca Raton, Florida, 29th edition.

[2] Milne-Thomson, L. (1970). Elliptic integrals. In Abramowitz, M. and Stegun, I. A., editors, *Handbook of Mathematical Functions with Formulas, Graphs, and Mathematical Tables*. U.S. Government Printing Office, Washington, DC, ninth edition.

[3] Pachter, M. et al. (2001). Minimizing radar exposure in air vehicle path planning. In *Proceedings of the 41st Israel Annual Conference on Aerospace Sciences*, Tel-Aviv, Israel.

[4] Skolnik, M. I., editor (1990). *Radar Handbook*. McGraw-Hill, New York, New York, second edition.

[5] Visual Numerics, I. (1997). *IMSL Math Library*, volume 1. Visual Numerics, Inc.

Chapter 11

ROBUST RECURSIVE BAYESIAN ESTIMATION AND QUANTUM MINIMAX STRATEGIES

P. Pardalos
Center for Applied Optimization,
Dept. of Industrial and Systems Engineering, University of Florida
pardalos@ufl.edu

V. Yatsenko
Scientific Foundation of Researchers and Specialists
on Molecular Cybernetics and Informatics
yatsenko@ise.ufl.edu

S. Butenko
Dept. of Industrial and Systems Engineering,
University of Florida
butenko@ufl.edu

Abstract The problem of a recursive realization of Bayesian estimation for incomplete experimental data is considered. A differential-geometric structure of nonlinear estimation is studied. It is shown that the use of a rationally chosen description of the true posterior density produces a geometrical structure defined on the family of possible posteriors. Pythagorean-like relations valid for probability distributions are presented and their importance for estimation under reduced data is indicated. A robust algorithm for estimation of unknown parameters is proposed, which is based on a quantum implementation of the Bayesian estimation procedure.

  Key words: Bayesian estimation, robust estimation, control, optimization, quantum detection, model approximation

R. Murphey and P.M. Pardalos (eds.), Cooperative Control and Optimization, 213–232.
© 2002 *Kluwer Academic Publishers.*

1.   Introduction

At present, different principles and methods are used to estimate characteristics of complex controlled objects [12, 15, 16]. However, in many cases these estimates require a large amount of experimental data, which are not always available. Therefore, it is important to formulate and solve the problems concerned with estimation of the object states, when the information about the object is incomplete. In the present paper it is proposed to apply a procedure, based on a recursive nonlinear estimation of the object parameters [11]. Some geometric aspects of robust Bayesian estimation are examined, and the possibility to build a quantum implementation of the minimax strategy is discussed.

Consider a stochastic system in which data are observed at discrete time moments $t = 1, 2, \ldots$. Let the dependence of the system output y_t on previous data be described by a suitable parametric family of probability densities

$$p(y_t \mid u_t, x_{t-1}, \theta) = m_t(\theta), \qquad (1)$$

conditioned on the latest input u_t, the data $x_{t-1} = \{(y_r, u_r), r = 1, \ldots, t-1\}$ observed up to the time $t - 1$, and constant *parameters* $\theta \in \Theta$. The short notation $m_t(\theta)$ will be preferred when (1) is taken as a function of the unknown parameter θ for the known data.

Let the prior uncertainty of θ be expressed through a probability density $p_0(\theta)$ built on the basis of available initial information. Then the update of the density $p_{t-1}(\theta)$ by new data is defined by the Bayes theorem. If the input generator employs no other information about θ than observed data [13],

$$p(y_t \mid u_t \, x_{t-1}, \theta) = p(y_t \mid u_t \, x_{t-1}), \qquad (2)$$

then the Bayes rule simplifies to the multiplicative "composition law"

$$p_t(\theta) = \alpha p_{t-1}(\theta) m_t(\theta), \qquad (3)$$

where α is the normalization factor.

Note that instead of trying to estimate directly the unknown parameters θ, we prefer to gather complete information about them expressed through the posterior density $p_t(\theta)$. This standpoint has strong conceptual advantages [13], but the construction of the posterior density according to the functional relation (3) is rarely feasible. The main difficulty is that the dimension of the statistic sufficient to determine the true posterior $p_t(\theta)$ may be too large or even permanently growing with the increasing number of the observed data. Hence, with real computing facilities, we are forced to use a reduced statistic instead of the sufficient one.

2. Differential geometry of Bayesian estimation

In this section we review some basic results and definitions which are used in this paper. We prefer intuition, rather than being rigorous. For a full-detail description the reader is referred to standard textbooks [5, 10]. The statistical connections are elaborated in [6, 11].

Differentiable manifold. The *differentiable manifold P* is a space (with some suitable topological properties) whose points can be mapped, at least locally, to a finite-dimensional Euclidean space. Any suitable mapping (homeomorphism) $\psi : P \to R^{\dim P}$ can stand for a coordinate function. The transformations from one coordinate function to another are assumed to be differentiable and regular (i.e. diffeomorphisms).

Lie group. The *Lie group M* is a group and a differentiable manifold at the same time. The group operations (composition and inversion of group elements) are assumed differentiable. Being a manifold, M can be provided with a coordinate function $\Phi : M \to R^{\dim M}$.

Homomorphism of Lie groups. A mapping $\chi : M \to X$ from a Lie group M to another Lie group X is called a *homomorphism* of Lie groups if it is an algebraic homeomorphism that retains group operations and consequently, neutral elements:

(a) $\chi(mm') = \chi(m)\chi(m')$ for any $m, m' \in M$;

(b) $\chi(m^{-1}) = \chi^{-1}(m)$ for any $m \in M$;

(c) $\chi(1_M) = 1_X$;

and, at the same time, it is a differentiable mapping. The *rank* of a homomorphism is defined as the rank of the corresponding Jacobian matrix.

Action of a Lie group on a manifold. We say that a Lie group M acts on a differentiable manifold P (on the right) if there exists a mapping $\beta : P \times M \to P$ satisfying the conditions

(a) if $1_M \in M$ is the neutral element of M, then $\beta(p, 1_M) = p$ for all $p \in P$;

(b) if $m, m' \in M$, then $\beta(\beta(p, m), m') = \beta(p, mm')$. The action is called *free* if $\beta(p, m) = p$ implies $m = 1_M$.

Fibre bundle. Let P be a differentiable manifold and M a Lie group acting on P (on the right). M induces an equivalence relation on P; p, p' are equivalent if for some $m \in M$ we have $p' = pm$. Denote by P/M the quotient space of P relative to this equivalence and assume that the action of M on P is free, the canonical projection $\pi : P \to P/M$ is differentiable and P is locally trivial (diffeomorphic to the product manifold $P/M \times M$). Then we say that P together with the action of M on P forms a *principal fibre bundle over* P/M, denoted by $P(P/M, M)$.

Tangent space of a manifold. The tangent space of a differentiable manifold P at the point $p \in P$ is a vector space obtained by local linearization
around p. It is composed of tangent vectors of the smooth curves passing
through p. The *natural basis* of the tangent space T_ω, associated with
specific coordinates $\omega = \psi(p)$, is formed by the partial derivatives with
respect to ω_i, namely $\frac{\partial}{\partial \omega_i}, i = 1, 2, \ldots \dim P$, which applied to any differentiable function on P generates directional derivatives along particular
coordinate curves.

Metric tensor. We define the inner product of two basis vectors $\frac{\partial}{\partial \omega_i}$,
and $\frac{\partial}{\partial \omega_j}$ at a generic point $p(\theta; \omega) \in P$ with the coordinates ω as

$$g_{ij}(\omega) = \left\langle \frac{\partial}{\partial \omega_i}, \frac{\partial}{\partial \omega_j} \right\rangle = E_p \left[\left(\frac{\partial}{\partial \omega_i} \log p(\theta; \omega) \right) \left(\frac{\partial}{\partial \omega_j} \log p(\theta; \omega) \right) \right].$$

(4)

Expectation is always taken relative to the point in which the tangent
space is constructed. The quantities $g_{ij}(\omega)$, $i, j = 1, 2, \ldots, \dim P$ form
together the so-called *metric tensor*. The corresponding matrix (g_{ij})
is a Bayes variant of the well-known Fisher information matrix. This
definition makes the metric tensor invariant with respect to coordinate
transformations both in θ and in ω [1, 2].

Orthogonality. Using (4), the inner product of two vectors $A = \sum_{i=1}^{\dim P} A^i \frac{\partial}{\partial \omega_i}$,
$B = \sum_{i=1}^{\dim P} B^i \frac{\partial}{\partial \omega_i}$ can be expressed as

$$\langle A, B \rangle = \sum_{i=1}^{\dim P} \sum_{j=1}^{\dim P} A^i g_{ij} B^j.$$

We call the tangent vectors A, B *orthogonal* if $\langle A, B \rangle = 0$.

We adopt two main assumptions:
(A1) A fixed amount of data is sufficient to evaluate $m_t(\theta)$.
(A2) The functions $m_t(\theta)$, $t = 1, 2, \ldots$ are nowhere vanishing.
Assumption (A1) excludes from our investigation a problem where the
filtering of data is necessary (e.g. estimation of ARMAX model parameters). Assumption (A2) restricts the class of possible models. Typically,
it eliminates regression models with bounded (e.g. uniformly distributed)
noise.

Under these assumptions, we examine recursive Bayesian estimation
using only a reduced description of the posterior density.

The main findings include [12]:

1) The family of possible posterior densities, called posterior family, can be imbedded into an *exponential family*.

2) There exists a reduced statistical description of the true posterior density, which is Bayes-closed in the sense of a group homomorphism.

3) The Bayes-closed description can be reduced to a collection of equivalence classes of densities corresponding to some values of the used parameters.

4) Any of the equivalence classes represents an *exponential subfamily* of the posterior exponential family. All these exponential subfamilies have the same "generating" functions.

5) There exist orthogonal parameters, one of which identifies the posterior density accounts for the transition within a specific equivalence class, while the other one reflects a movement across particular equivalence class which is *orthogonal* at any point to the appropriate equivalence class.

6) The existence of the orthogonal parametrization implies that it is possible to construct a *curved exponential subfamily* of the posterior exponential family which is orthogonal to the equivalence classes that it intersects.

7) The orthogonal setting of the curved exponential subfamily relative to the equivalence class enriches the intersection point with important *extremal properties*: this point minimizes the Kullback – Leibler distance from any point of the curved exponential subfamily and, at the same time, the dual Kullback – Leibler distance from the true posterior density [12].

3.   Optimal recursive estimation

Recall that we are concerned with recursive discrete-time Bayesian parameter estimation for the model (1) with the following assumptions.

(A1) The data statistics sufficient to determine $m_t(\theta)$ as a function of the unknown parameters θ has a limited finite dimension at any t.

(A2) The function $m_t(\theta), t = 1, 2, \ldots$ have the same support Θ, i.e. $m_t(\theta) > 0$ almost surely for any $\theta \in \Theta$.

Under these assumptions, we can imbed the Bayes estimation problem into a standard differential-geometric model. Because of the assumption

(A2), the Bayes rule (3) can be converted into the form [3]

$$\log p_t(\theta) = \text{const} + \log p_0(\theta) + \sum_{r=1}^{t} \log m_r(\theta).$$

The additive composition law inspires us to imbed the term $\sum_{r=1}^{t} \log m_r(\theta)$ into a vector space, say F. In such case, some points of F coincide with the log-likelihoods $\log p(y|u, x, \theta)$ for different data, other points result from their additive composition and are added just to complete a vector space. Clearly, this way of composition is by far not a unique one, but it is the simplest one in terms of both its definition and dimensionality of the enveloping vector space.

The "extended" space of likelihoods $M = \exp F$ has a useful property of being a group with the group operation defined by pointwise multiplication of functions. Due to an obvious bijection $m(\theta) \leftrightarrow f(\theta) = \log m(\theta)$, if $f = \log m$, $f' = \log m'$ then $mm' = \exp|f + f'|$, $m^{-1} = \exp|-f|$. Note also that $1 = \exp 0$.

As the posterior density is normalized in order to give a unit integral over Θ, there is no need to care about the norm of the likelihood functions $m(\theta)$ during computations. This is why we consider equivalence classes $\tilde{m}$ of functions $m(\theta)$ differing only by their norm. The same equivalence relation will also be applied to functions $q(\theta)$ representing unnormalized posterior densities.

These considerations motivate the following definition.

Definition 3.1 *Let Φ be the set composed of log-likelihood functions $\log p(y|u, x, \theta)$ corresponding to all possible values of data (y, u, x). Denote by F the linear hull of Φ — the smallest vector subspace of the enveloping vector space L that contains Φ.*

Introduce the space of unnormalized densities $Q = \exp L$ and likelihoods $M = \exp F$ as the spaces of exponential functions of the elements of L and F, respectively.

Define an equivalence relation $\sim$ on the vector space L so that two functions from L are equivalent if they differ at most by an additive finite constant. This relation induces on M and Q the equivalence of proportionality — equality up to a positive finite multiplicative factor.

By the geometric model of Bayesian estimation we will call the object $(\tilde{Q}, \tilde{M}, \beta)$ with the following meaning: $\tilde{Q}$ and $\tilde{M}$ are quotient spaces $Q/\sim$ and $M/\sim$ respectively, and the mapping $\beta : \tilde{Q} \times \tilde{M} \to \tilde{Q}$ is defined by

$$\tilde{q}\tilde{m} = \beta(\tilde{q}, \tilde{m}) = cq(\theta)m(\theta) \,|\, c > 0, \ q \in \tilde{q}, m \in \tilde{m} \qquad (5)$$

The mapping β is the unnormalized version of the Bayes rule (3).

To avoid technical subtleties with infinitely-dimensional spaces, we adopt the following assumption:

(A3) The linear hull F of the set Φ of log-likelihood functions $\log p(y|u, x, \theta)$ over all possible data (y, u, x) has a finite dimension N.

Finally, in the sequel we restrict our attention to the case:

(A4) Let $\widetilde{Q} = \widetilde{q_0 M} = \left\{ \widetilde{q_0 m} | \widetilde{m} \in \widetilde{M} \right\}$ for an arbitrary but fixed prior density $q_0 \in \widetilde{q}_0$.

It can be verified that the above geometric model is well defined in the sense of differential geometry:

Lemma 3.1 *Assume (A1)–(A4) are fulfilled. Then $\widetilde{Q}$ is a differentiable manifold, $\widetilde{M}$ is a Lie group with the group operation induced by pointwise multiplication of functions, and β defined by (5) specifies the action of $\widetilde{M}$ on $\widetilde{Q}$.*

To be able to introduce the Riemannian geometry for the Bayesian estimation problem, we introduce in addition to $\widetilde{Q}$ the corresponding space of normalized densities.

Definition 3.2 *Let $\widetilde{P} \subset \widetilde{Q}$ contain all of the equivalence classes $\widetilde{q} = \{cq(\theta)|c > C\}$ from $\widetilde{Q}$ which are composed of integrable functions for which $\int q(\theta)d\lambda < \infty$. By the posterior family P we mean the space of normalized densities $p(\theta)$ corresponding in a one-to-one way to particular equivalence classes $\widetilde{p} \in \widetilde{P}$*

$$p(\theta) = \frac{q(\theta)}{\int q(\theta)d\lambda(\theta)}$$

The space of normalized densities is also well defined from a differential-geometric viewpoint [6, 1, 2].

Lemma 3.2 *Assume (A1)–(A4) are fulfilled. Then P is a differentiable manifold.*

Now we are ready to formulate the problem. According to (A3), a complete description of $\widetilde{m} \in \widetilde{M}$ can be done through the coordinate function $\Phi : \widetilde{M} \to R^N$. Note that Φ can be given other interpretations. From the statistical viewpoint, Φ is a statistics, i.e. a function of observed data. From the hierarchical Bayesian viewpoint, Φ is a hyper-parameter that specifies uniquely the posterior density of the unknown parameters θ (provided that the prior density is fixed).

In a concrete implementation of recursive Bayesian estimation we are often forced to store just a reduced description χ of $\tilde{m}$ whose dimension is smaller than N. Choosing an inappropriate description χ, we can completely loose control over estimation. Thus, we are searching for a "good" description which would retain basic operations over $\widetilde{M}$.

Optimal recursive estimation.   Find a homomorphism (of Lie groups) of rank $n < N$

$$\chi : \widetilde{M} \to R^n. \tag{6}$$

If such mapping exists, perform the optimal recursive estimation on equivalence classes $[\tilde{q}] = \tilde{q}_0[\tilde{m}]$ with

$$[\tilde{m}] = \left\{ \tilde{m}' \in \widetilde{M} \mid \chi(\tilde{m}') = \chi(\tilde{m}) \right\} \tag{7}$$

in the sense that

$$[\tilde{q}_{t-1}\tilde{m}_t] = [\tilde{q}_{t-1}][\tilde{m}_t] = \{\tilde{q}'_{t-1}\tilde{m}'_t \mid \tilde{q}'_{t-1} \in [\tilde{q}_{t-1}, \tilde{m}'_t \in [\tilde{m}_t]\} \tag{8}$$

It is the relation (8) that has motivated us to search for a description of χ which would have the basic property (a) of homomorphisms (see Section 2). If a homomorphic description χ is applied, there is no difference between the one-shot and recursive estimation. This is why we call such a description Bayes-closed.

Clearly, using a homomorphism $\chi : \widetilde{M} \to R^n$ of rank $n < N$ we are not yet able to specify the true density $p_t \in P$ exactly, but we can determine the equivalence class

$$[p_t] = \left\{ p \in P \mid p(\theta) = \frac{p_0(\theta)m(\theta)}{\int p_0(\theta)m(\theta)d\lambda(\theta)}, m \in \tilde{m}, \tilde{m} \in [\tilde{m}_t] \right\} \tag{9}$$

in which p_t lies. Thus, optimal recursive estimation implies a closely related problem of "approximation" — to find a rationally justified representative of the equivalence class $[p_t]$ which could be used in a subsequent decision making. Our aim is to suggest a procedure that would give the same result both for the case of p_t known and the case when just $\chi \left(\prod_{r=1}^t \tilde{m}_r \right)$ is given (in addition to knowledge of p_0).

Approximation of recursive estimation.   Given a homomorphism $\chi : \widetilde{M} \to R^n$ of rank n, find a subfamily $Z \subset P$ orthogonal to all equivalence classes $[p] \subset P$ defined by (9). If such a family exists, construct the approximating density $\tilde{p}_t$ by the orthogonal projection of the true posterior p_t into Z.

We notice that the manifold P specified by Definition 3.2 represents the exponential family with a generic point written in local coordinates $\omega = (\omega^1, \ldots, \omega^N)$ as

$$p(\theta; \omega) = \alpha p_0(\theta) \exp\left[\sum_k 1^N \omega^k f_k(\theta) \right]. \tag{10}$$

This observation follows directly from Definition 3.1, since the set

$$\{f_1, \ f_2, \ \ldots, \ f_N\}$$

forms a basis of the vector space F.

Given a basis $f_1, \ldots, f_N$, $\tilde{m} \in \widetilde{M}$ can be described completely by local coordinates $\omega^1, \ldots, \omega^N$. The construction of a reduced but homomorphic description of $\tilde{m}$ is described by the following proposition.

Proposition 3 *Define the mapping* $\chi : \widetilde{M} \to R^n$ *of rank* n *through*

$$\chi^t(\tilde{m}) = E_{p_0}[h_i(\theta) \log m(\theta)], \tag{11}$$

where $h_i(\theta) \in F$ *are* n *linearly independent functions such that* $E_{p_0}[h_i] = 0$, $i = 1, \ldots, n$ *and* $m(\theta)$ *stands for an arbitrary representative of the equivalence class* $\tilde{m}$. *The mapping* (11) *forms a homomorphism (of Lie groups). Any other homomorphism is related to* (11) *by an isomorphism.*

Proof. Any (closed regular) Lie subgroup $\widetilde{M}_v \subset \widetilde{M}$ is a kernel of some homomorphism of $\widetilde{M}_v$ (see Theorem 19 in [2] and Theorem 6.14 in [5]). One obvious homomorphism with the kernel $\widetilde{M}_v$ is the canonical homomorphism $\chi^* : \widetilde{M} \to \widetilde{M}/\widetilde{M}_v$ which assigns to any $\tilde{m}$ the corresponding equivalence class $\tilde{m}\widetilde{M}_v$. Any other homomorphism is related to the canonical one through an isomorphism to $\widetilde{M}/\widetilde{M}_v$.

Therefore, it is sufficient to find a definition of the mapping χ that ensures that $\chi(\tilde{m}\tilde{m}_v) = \chi(\tilde{m})$ for any $\tilde{m}_v \in \widetilde{M}_v$, $\tilde{m} \in \widetilde{M}$ and, at the same time, $\chi(\tilde{m}) \neq \chi(\tilde{m}')$ if $\tilde{m}\widetilde{M}_v \neq \tilde{m}\widetilde{M}_v$. Taking into account the analytic expression for fibres (13), it is easy to verify that the definition (11) fulfills the above condition in the case when the terms $E_{p_0}[h_i f_i]$ and $E_{p_0}[h_i]$ are equal to zero. ∎

The kernel of the homomorphism χ defined by (11) is a Lie subgroup $\widetilde{M}_v$ of $\widetilde{M}$ [5]. The use of a reduced description χ induces an equivalence relation on $\widetilde{Q}$; namely, $\tilde{q}, \tilde{q}'$ are equivalent if $\tilde{q}' = \tilde{q}\tilde{m}_v$ for some $\tilde{m}_v \in \widetilde{M}_v$. Thus, we are able to distinguish only among the sets

$$\tilde{q}\widetilde{M}_v = \left\{ \tilde{q}\tilde{m}_v \mid \tilde{m}_v \in \widetilde{M}_v \right\} = \{ \tilde{q}\tilde{m} \mid \chi(\tilde{m}) = 0 \}, \tag{12}$$

which represent the equivalence classes $[\tilde{q}] = \tilde{q}[\tilde{m}]$. In terms of differential geometry, the action of $\widetilde{M}_v$ on $\tilde{Q}$ produces a principal fibre bundle $\tilde{Q}(\tilde{Q}/\widetilde{M}_v, \widetilde{M}_v)$. Thus, the equivalence classes $[\tilde{q}]$ represent submanifolds of $\tilde{Q}$ which are mapped, in a straightforward way, to the equivalence classes of normalized densities $[p] \subset P$.

The Lie subgroup $\widetilde{M}_v$ (the kernel of χ) corresponds to a vector subspace F_v, through the formula $\widetilde{M}_v = (\exp F_v)/ \sim$. Using (11) implies that F_v is spanned by the functions $v_1(\theta), \ldots, v_{N-n}(\theta)$ which satisfy $E_{p_0}[h_i v_j] = 0$, $i = 1, \ldots, n$, $j = 1, \ldots, N-n$. A generic density of the equivalence class (fibre) $[p]$ through p can be expressed in local coordinates $r = (r^1, \ldots, r^{N-n})$ as

$$p(\theta; s, r) = \alpha p(\theta) \exp\left[\sum_{j=1}^{N-n} r^j v_j(\theta)\right]. \tag{13}$$

Hence, fibres form the exponential subfamilies of P and the posterior exponential family may be written in the form

$$p(\theta; s, r) = \alpha p_0(\theta) \exp\left[\sum_{i=1}^{n} s^i h_i(\theta)\right] \exp\left[\sum_{j=1}^{N-n} r^j v_j(\theta)\right]. \tag{14}$$

Re-parametrize (14) as follows

$$p(\theta; \bar{s}, r) = \alpha p_0(\theta) \exp\left[\sum_{k=1}^{N} \omega^k(\bar{s}) f_k(\theta)\right] \exp\left[\sum_{j=1}^{N-n} r^j v_j(\theta)\right], \tag{15}$$

where the first exponent models a movement of the point p over an n-dimensional submanifold (subsurface) of P (see 10) through a smooth vector-valued function $\omega = \omega(\bar{s})$ of $\bar{s} = (\bar{s}^1, \ldots, \bar{s}^n)$, while the second one explains changes within the corresponding fibre. The following proposition solves the problem of finding a parametrization $(\bar{s}, r)$, such that at any point p of

$$Z = \left\{ p \in P \mid p(\theta; \bar{s}) = \alpha p^*(\theta) \exp\left[\sum_{k=1}^{N} \omega^k(\bar{s}) f_k(\theta)\right] \right\}, \tag{16}$$

the tangent vectors of the $(N-n)$-dimensional fibre $[p]$ in p and of the n-dimensional family Z (16) in a reference point p^n are orthogonal:

$$\left\langle \frac{\partial}{\partial \bar{s}^i}, \frac{\partial}{\partial r^j} \right\rangle = 0. \tag{17}$$

Combining (4) and (15), this condition reads

$$E_p \left[\left(\frac{\partial}{\partial \bar{s}^i} log\, p(\theta; \bar{s}, r) \right) \left(\frac{\partial}{\partial r^j} log\, p(\theta; \bar{s}, r) \right) \right] = 0. \qquad (18)$$

Proposition 4 ([11]) *Given a set of equivalence classes* $[p] \subset P$ *produced by the Lie subgroup* $\widetilde{M_0} = \ker \chi$, *there exists a submanifold* $Z \subset P$ *through a given point* $p^n \in P$ *that is orthogonal to all equivalence classes which it intersects. An intersection point* $\hat{p} \in [p] \cap Z$ *fulfills the identities*

$$E_{p_0}[h_i log\, \hat{p}] = E_{p_0}[h_i log\, p] \qquad (19)$$

$$E_{\hat{p}}[v_j] = E_{p^*}[v_j]. \qquad (20)$$

Proof. After taking the partial derivative in the first term of (18), we get

$$\int \left(\frac{\partial}{\partial \bar{s}^i} p(\theta; \bar{s}, r) \right) \left(\frac{\partial}{\partial \bar{r}^i} log\, p(\theta; \bar{s}, r) \right) d\lambda(\theta) = 0.$$

By substituting (15) for $p(\theta; \bar{s}, r)$ in the second term and taking into account that the densities are normalized, we derive after some manipulations

$$\int \left(\frac{\partial}{\partial \bar{s}^i} p(\theta; \bar{s}, r) \right) v_j(\theta) d\lambda(\theta) = 0.$$

Finally, changing the order of integration and partial derivative $\frac{\partial}{\partial \bar{s}^i}$, we derive $\frac{\partial}{\partial \bar{s}^i} E_p[v_j] = 0$. Thus, $E_p[v_j] = const$ for any $p \in Z$ which implies directly (20). The identity (19) results from the fact $\hat{p} \in [p]$ and the definition (11) of χ. ∎

Proposition 4 implies that there exists an orthogonal parametrization $(\bar{s}, r)$ of the manifold P making the changes within Z and within equivalence classes $[p]$ locally independent — in the sense of a zero inner product of the tangent vectors.

Z can be interpreted as follows. Define at each point $(\bar{s}, r) \in P$ a vector space $T(Z)$ consisting of the vectors that are tangential to P and, at the same time, orthogonal to the tangent space of $[p]$. Then the submanifold Z such that its tangent space coincide with $T(Z)$ at any point is called the *integral submanifold* of the vector fields determined by the basis vectors of $F(Z)$.

In a statistical context, the family Z is said to be a curved exponential family [1, 2]. Geometrically speaking, Z is an n-dimensional differentiable submanifold imbedded in the N-dimensional manifold P of the posterior exponential family.

The relation (20) of Proposition 4 implies that $E_p[v_j] = E_{p'}[v_j] = E_{p^n}[v_j]$ for any $p, p' \in Z$ and, hence $E_{\alpha p + (1-\alpha)p'}[v_j] = E_p[v_j]$ for $0 \leq$

$\alpha \leq 1$. Thus, Z can also be interpreted as a differentiable submanifold imbedded in a mixture Z^t that is defined as the smallest space of densities which contains Z and is closed with respect to a convex composition $\alpha p + (1-\alpha)p'$, $0 \leq \alpha \leq 1$. It can be proved (see the proof of Proposition 4) that Z is orthogonal to all fibres $[p]$ that it intersects.

The optimal local properties of orthogonal parametrization are accompanied by very attractive global properties.

Corollary 3.1 *The intersection point* $\widehat{p} \in [p] \cap Z$, $Z \ni p^*$ *represents a unique solution of the following dual optimization problems*

$$\min_{\widehat{p} \in Z} \int \widehat{p}(\theta) \log \frac{\widehat{p}(\theta)}{p(\theta)} d\lambda(\theta), \tag{21}$$

$$\min_{\widehat{p} \in [p]} \int p^*(\theta) \log \frac{p^*(\theta)}{\widehat{p}(\theta)} d\lambda(\theta). \tag{22}$$

Proof. The corollary follows by application of Theorems 3.8 and 3.9 in [13]. For more detail on "projection geometry" see [11]. ∎

The above formulas have a great intuitive appeal. First, $\widehat{p}$ represents a point of $[p]$ that minimizes the Kullback – Leibler distance from a reference density p^* (typically the prior density p_0). At the same time, $\widehat{p}$ represents the point of Z that minimizes the dual Kullback – Leibler distance from the true density p (although we do not know p completely!).

We summarize the previously discussed results in an algorithmic form.

The Algorithm

Initialization

(I1) Choose a prior $p_0(\theta)$.

(I2) Specify the posterior exponential family P (see (10)).

(I3) To determine the form (14) of the posterior exponential family, choose a base $\{h_1, \ldots, h_n, v_1, \ldots, v_{N-n}\}$ of the vector space F that fulfills $E_{p_0}[h_i] = 0$, $E_{p_0}[v_i] = 0$, $E_{p_0}[h_i v_i] = 0$ for all $i = 1, \ldots, n$, $j = 1, \ldots, N - n$.

(I4) Evaluate the metric tensor at p_0

$$g_{ij}(p_0) = \langle \frac{\partial}{\partial s_i}, \frac{\partial}{\partial s_j} \rangle_{p_0} = E_{p_0}[h_i h_j]$$

for $i, j = 1, \ldots, n$, and its inversion $g^{-1}(p_0)$.

(P5) Set $\chi_0 = 0$ and $t := 1$.

Recursive steps

(R1) Observe new data u_t, y_t.

(R2) Evaluate

$$\Delta \chi_t' = E_{p_0}[h_i(\theta) \log n_t(\theta)].$$

(R3) Update

$$\chi_t' = \chi_{t-1}' + \Delta \chi_t'.$$

(R4) Determine s_t, such that

$$s_t = g^{-1}(p_0)\chi_t = s_{t-1} + g^{-1}(p_0)\Delta \chi_t.$$

(R5) For a prespecified (possibly time variable) reference density p_t^*, determine r_t, such that

$$E_{\widehat{p}_t}[v_j(\theta)] = E_{p_t^*}[v_j(\theta)].$$

(R6) Increment $t := t + 1$ and go to step (R1).

4. Quantum realization of minimax Bayes strategies

The quantum realization of optimal strategies [8, 16] is very attractive for use in practical design problems of safety analysis systems because it can predict an ultimate performance of complex controlled objects.

The applications of the quantum detection theory to the practical signal sets have been carried out by many researchers. It has been revealed that the quantum Bayes formula is too complex to obtain the analytical solutions except for a few quantum signal sets, while the quantum minimax can be solved for more signal sets. We can confirm the correctness and effectiveness of the quantum minimax formula for binary phase shift keyed, ternary symmetric, and 3 phase shift keyed coherent state signal sets.

Although the analytical solutions of the optimum detection operators are purely mathematical descriptions of the quantum detection process, they provide important information for the implementation of the problem of designing a practical system with the minimum error, the so-called "optimum quantum receiver". In this section, we discuss the advantages of the quantum minimax approach with respect to the quantum Bayes formula. The numerical analysis is applied for this purpose. In the case of ternary symmetric signal set, we can show the precise error probability by the quantum minimax formula including the former one given in [4].

This section is devoted to a brief summary of the quantum detection theory, which offers fundamental formulas of the quantum Bayes and minimax strategies. There are some strategies in the quantum detection theory based on the concept similar to the classical decision theory. That is, the quantum Bayes, Neyman-Pearson, and minimax strategies are considered. In this section, the quantum Bayes and minimax are investigated because the quantum Bayes strategy gives the minimum risk when the signal prior-probabilities are known at a detection side. On the other hand, when the signal prior-probabilities are unknown at a detection side, the quantum minimax strategy is the most useful since it guarantees the minimum worst risk for all signal prior-probabilities. This risk is equal to the minimum risk achieved by the quantum Bayes with the least favorable signal prior-probabilities. Furthermore the quantum minimax can simplify the derivation process of a solution sufficiently comparing to the case of the quantum Bayes.

Since we discuss the application of the quantum detection theory to safety analysis systems, the risk function to be optimized is the error probability defined as follows:

$$Pe = 1 - \sum_{i=1}^{M} \gamma_i Tr \, \widehat{\rho}_i \widehat{D}_i, \qquad (23)$$

where $\widehat{\rho}_i$, $\widehat{D}_i$ and γ_i are the density operator for a signal quantum state, the detection operator, and the prior-probability of a signal respectively, and M is the number of signals. The quantum detection corresponds to a separation of the signal space which is a subspace of the Hilbert space spanned by signal quantum states $\{\widehat{\rho}_i\}$. Therefore, the detection operators $\{\widehat{D}_i\}$ are the solutions of the identity

$$\sum_{i=1}^{M} \widehat{D}_i, = \widehat{I}, \qquad (24)$$

Since we assume that signal quantum states are pure states for practical calculation, the following lemma may be useful.

Lemma 4.1 ([3]) *When the signal quantum states are linearly independent, the minimum of the error probability is indeed projection-valued.*

Namely, the detection operators can be described by the measurement states $|\omega_i\rangle$, as follows:

$$\widehat{D}_i, = |\omega_i\rangle\langle\omega_i|, \quad \text{and} \quad \langle\omega_i|\omega_j\rangle = \delta_{ij}, \qquad (25)$$

so that the signal quantum states $\{|\omega_i\rangle\}$, can be defined by measurement states with variables $y_{ji} \equiv \langle \omega_j | \psi_i \rangle$

$$|\psi_i\rangle = \sum_{j=1}^{M} y_{ij}|\omega_j\rangle. \tag{26}$$

Since this equation can be represented in the matrix form as

$$\begin{bmatrix} |\psi_1\rangle \\ \vdots \\ |\psi_M\rangle \end{bmatrix} = \begin{bmatrix} y_{11} & \cdots & y_{M1} \\ \vdots & \ddots & \vdots \\ y_{1M} & \cdots & y_{MM} \end{bmatrix} \cdot \begin{bmatrix} |\omega_1\rangle \\ \vdots \\ |\omega_M\rangle \end{bmatrix}, \tag{27}$$

the measurement states can be explained as follows:

$$\begin{bmatrix} |\omega_1\rangle \\ \vdots \\ |\omega_M\rangle \end{bmatrix} = \begin{bmatrix} y_{11} & \cdots & y_{M1} \\ \vdots & \ddots & \vdots \\ y_{1M} & \cdots & y_{MM} \end{bmatrix}^{-1} \cdot \begin{bmatrix} |\psi_1\rangle \\ \vdots \\ |\psi_M\rangle \end{bmatrix}. \tag{28}$$

In addition, the error probability is also written using y_{ji}

$$Pe = 1 - \sum_{i=1}^{M} \gamma_i |y_{ii}|^2. \tag{29}$$

Therefore we need to find the matrix $\{y_{ji}\}$ in order to minimize the error probability and to determine the optimum measurement states. The rules to find the optimum matrix are explained in the following.

The quantum Bayes strategy is the minimization of the error probability with respect to the detection operators

$$\min_{\widehat{D}} Pe = \min_{\widehat{D}} \left\{ 1 - \sum_{i=1}^{M} \gamma_i Tr \, \widehat{\rho}_i \widehat{D}_i \right\}. \tag{30}$$

The necessary and sufficient conditions for the optimum detection operators become

$$\widehat{D}_j [\gamma_j \widehat{\rho}_j - \gamma_i \widehat{\rho}_i] \, \widehat{D}_i = 0, \quad \forall i, j, \tag{31}$$

$$\widehat{G} - \gamma_i \widehat{\rho}_i \widehat{D}_i = \sum_{i=1}^{M} \gamma_i \widehat{D}_i \widehat{\rho}_i, \tag{32}$$

where

$$\widehat{G} = \sum_{i=1}^{M} \gamma_i \widehat{\rho} \widehat{D}_i = \sum_{i=1}^{M} \gamma_i \widehat{D}_i \widehat{\rho}. \tag{33}$$

By means of the last lemma, (31) can be expressed through y_{ij} as follows:

$$\gamma_k y_{kk} y_{mk}^* - \gamma_m y_{km} y_{mm}^* = 0. \tag{34}$$

Since the inner products of the signal quantum states are related with y_{ji} as

$$k_{ij} = \langle \psi_i | \psi_j \rangle = \sum_{k=1}^{M} \langle \psi_i | \omega_k \rangle \langle \omega_k | \psi_j \rangle = \sum_{k=1}^{M} y_{kj} y_{ki}^*, \tag{35}$$

M^2 equations are given by (34) and (35) for M^2 unknown variables $\{y_{ji}\}$. Therefore all of $\{y_{ji}\}$ can be found from these equations.

The quantum minimax strategy is at the same time an application of the classical one, where the error probability is maximized with respect to the prior-probability, γ_i, and minimized by choosing the detection operators, $\widehat{D}_i$.

$$\min_{\widehat{D}} \max_{\gamma} Pe = \max_{\gamma} \min_{\widehat{D}} Pe. \tag{36}$$

The mathematical foundations of the quantum minimax detection rule such as the complete class theorem, existence theorem, and other were clarified in [4] taking into account the following facts of classical theory:

1) Any minimax rule is Bayes relative to the least favorable prior probability.

2) The minimax rule has a smaller maximum average risk than any other. The authors of [4] gave the necessary and sufficient conditions of the quantum minimax detection operator by taking into account the following.

3) The Bayes decision rule whose conditional risk is constant is a minimax rule. This provides the following minimax condition to the quantum Bayes formula given in (30) and (31):

$$Tr\widehat{\rho}_i \widehat{D}_i = Tr\widehat{\rho}_j \widehat{D}_j, \quad \forall \, i, j. \tag{37}$$

Using the relation $y_{ji} = \langle \omega_j | \psi_i \rangle$, (35) is represented as follows:

$$|y_{ii}|^2 = |y_{jj}|^2, \quad \forall \, i, j. \tag{38}$$

This condition makes it much easier to solve the nonlinear equations given by (33) and (34).

5.   Concluding remarks

The analysis of the problem, concerned with estimation of a state of complex controlled objects for restricted experimantal data results in the fact, that it is possible to solve it on the basis of the strategy of recursive nonlinear Bayesian estimation. The analysis of the differential-geometric estimation scheme structure allows to generalize the approach to the broader class of complex controlled objects. Assuming, that the diagnostic system of estimation of a complex controlled object admits its optical realization, the possibility to implement the minimax estimation strategy is demonstrated. This implementation may be performed in the form of a quantum communication computation system.

The geometry of nonlinear estimation provides a new insight into the question about the choice of a suitable "approximation family" [11]. The requirement that the reduced statistic has to be a homomorphism implies that the equivalence classes (fibres) are exponential families. The integral manifolds orthogonal to all fibres are curved exponential families, that can be viewed as intersections of the posterior exponential family with a mixture family. Hence, the mixture family is the best candidate for the approximation family provided it is orthogonal to all fibres.

The correctness and effectiveness of the quantum minimax formula were discussed in some cases of coherent state signal sets. The quantum minimax formula may be used to obtain analytical solutions for signal sets without a special skill level. The analytically obtained optimum detection operators provide a sufficient information to the implementation problem of the optimal quantum estimation.

References

[1] Amari, S.-I., Differential geometry of curved exponential families — curvatures and information loss. *The Annals of Statistics*, 10(2), pp. 357–385, 1982.

[2] Amari, S.-I., Differential-Geometrical Methods in Statistics, Springer–Verlag, New York, 1985.

[3] Belavkin V. P., O.Hirota, and R.L.Hudson, Quantum communication and measurement, Plenum Press, 1996.

[4] C.Bendjaballah, O.Hirota, and S.Reynaud, Quantum Aspect of Optical Communications, Lecture Note in Physics, Springer, LNP-378, 1991.

[5] Boothby, W. M., An Introduction to Differential Manifolds and Riemannian Geometry, Academic Press, London, 1975.

[6] Chentsov, N. N., Statistical Decision Rules and Optimal Inference (in Russian), Nauka, Moscow, 1972. English translation: Translation of Mathematical Monographs, 53. AMS, Rhode Island, 1982.

[7] Golodnikov, A., P. Knopov, P. Pardalos and S. Uryasev, "Optimization in the space of distribution functions and applications in the Bayes Analysis". In "Probabilistic Constrained Optimization" (S. Uryasev, Editor), pp. 102-131, Kluwer Academic Publishers, 2000.

[8] Helstrom C.W., Quantum Detection and Estimation Theory, Academic Press, 1976.

[9] Insua, D. R. and F. Ruggeri, Eds., Robust Bayesian Analysis, Lecture Notes in Statistics, Springer Verlag, 2000.

[10] Kobayashi, S. and K. Nomixu, Foundations of Differential Geometry, Vol. I, Interscience Publishers, New York, 1963.

[11] Kulhavý, R., Recursive Nonlinear Estimation: a Geometric Approach, Springer, London, 1996.

[12] Kulhavý, R., Differential geometry of recursive nonlinear estimation. In Preprints of the 11th IFAC World Congress, Tallinn, Estonia. Vol. 3, pp. 113-118, 1990.

[13] Peterka, V., Bayesian approach to system identification. In P. Eyichoff (ed.), Trends and Progress in System Identification, Pergamon Press, Oxford, 1981.

[14] Savage, L. J., The Foundations of Statistics, Wiley, New York, 1954.

[15] Sorenson, H. W., Recursive estimation for nonlinear dynamic systems, in J. C. Spall (Ed.), Bayesian Analysis of Time Series and Dynamic Models, Marcel Dekker, New York, 1988.

[16] Yatsenko, V., T. Titarenko, Yu. Kolesnik, Identification of the non-Gaussian chaotic dynamics of the radioemission back scattering processes, Proc. Of the 10th IFAC Simposium on System Identification SYSID'94 — Kobenhaven, 4-6 July, 1994. — V. 1. — P. 313 – 317.

[17] Wets, R., Statistical estimation from an optimization viewpoint, Annals of Operations Research, 85, pp.79-101, 1999.

Chapter 12

COOPERATIVE CONTROL FOR AUTONOMOUS AIR VEHICLES

Kevin Passino[†], Marios Polycarpou[‡], David Jacques[*], Meir Pachter[*], Yang Liu[†], Yanli Yang[‡], Matt Flint[‡] and Michael Baum[†]

[†] *Department of Electrical Engineering, The Ohio State University*
2015 Neil Avenue, Columbus, OH 43210-1272, USA

[‡] *Dept. of Electrical and Computer Engineering and Computer Sciences*
University of Cincinnati, Cincinnati, OH 45221-0030, USA

[*] *Air Force Institute of Technology, AFIT/ENY*
Wright Patterson Air Force Base, OH 45433-7765, USA

Abstract   The main objective of this research is to develop and evaluate the performance of strategies for cooperative control of autonomous air vehicles that seek to gather information about a dynamic target environment, evade threats, and coordinate strikes against targets. The air vehicles are equipped with sensors to view a limited region of the environment they are visiting, and are able to communicate with one another to enable cooperation. They are assumed to have some "physical" limitations including possibly maneuverability limitations, fuel/time constraints and sensor range and accuracy. The developed cooperative search framework is based on two inter-dependent tasks: (i) on-line learning of the environment and storing of the information in the form of a "target search map"; and (ii) utilization of the target search map and other information to compute on-line a guidance trajectory for the vehicle to follow. We study the stability of vehicular swarms to try to understand what types of communications are needed to achieve cooperative search and engagement, and characteristics that affect swarm aggregation and disintegration. Finally, we explore the utility of using

R. Murphey and P.M. Pardalos (eds.), Cooperative Control and Optimization, 233–271.
© 2002 *Kluwer Academic Publishers.*

biomimicry of social foraging strategies to develop coordination strategies.

Keywords: cooperative control, autonomous air vehicles

1.   Introduction

1.1.   UAAV operational scenarios

Recent technological advances are making it possible to deploy multiple Uninhabited Autonomous Air Vehicles (UAAVs) that can gather information about a dynamic threat/target environment, evade threats, and coordinate strikes against targets (see Figure 12.1). Here, we show a section of land (the black regions indicate mountains/rugged terrain where targets/threats generally do not reside) where there are targets with certain priorities (e.g., set by the command center) and threats with different severities relative to their ability to eliminate the UAAV or other assets. A target may also be a threat, but there may be threats that are not high priority targets, or targets that are not severe threats to the UAAV (e.g., the autonomous munition is assumed small and expendable, so targets are not considered to be threats to the munition itself). Mobile target/threats are designated by showing their path of travel. There are also stationary targets/threats. There are some targets/threats that are in groups, while others are distant from a group. There are groups with certain types of compositions (e.g., some in a high threat/low target priority group that are protecting a low threat severity/high target priority site). The proposed formulation can also allow for negative target priorities, which may include undesirable targets such as a truckload of civilians (collateral damage). The UAAVs are designated with shaded triangles (the black one has extra UAAV supervisory responsibility), and the ellipses show the region where the sensor resources of the UAAVs are currently focused (and they can only "see" targets/threats within this ellipse). In our general formulation we assume that in addition to sensing and maneuvering capabilities, the UAAVs have an ability to strike a target with on-board munitions, or the UAAV is a munition itself. There are some communications that are allowed between UAAVs, and these are represented via the bidirectional dotted arrows; however, there are communication limitations (e.g. bandwidth, delays, or range constraints). The general problem is to coordinate the movement of the UAAVs so that they can evade threats and locate and destroy the highest priority targets, before they expend their own resources (e.g. fuel).

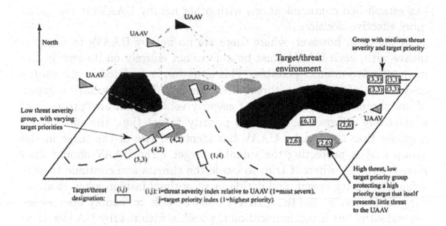

Figure 12.1. UAAVs operating in a dynamic target/threat environment.

1.2. UAAV cooperation to enhance mission effectiveness

While each UAAV could certainly operate independently of the others, the overall mission effectiveness can be improved via communications and coordination of their activities. For instance, for the situation depicted in Figure 12.1 suppose that one UAAV has identified the (2, 4)-target/threat (since it is in the focus of its sensors) and another UAAV has just identified a (4, 2)-target/threat. The notation used here is that (i, j) implies threat severity index i and target priority index j, where 1 denotes the highest priority (or the most severe threat). Further suppose that in the recent past the second UAAV also identified the (4, 2)-target/threat to its southwest. Note that this "identification" entails estimating the position of the target/threat, and the type of the target/threat (i.e., an estimate of the target/threat indices given in the figure), and there is uncertainty associated with each of these. In Figure 12.1 the "supervisory" UAAV is capable of receiving information from all other UAAV's subject to range limitations and data loss assumptions. This communication could be direct or via retransmission by all UAAV's. In this situation suppose the supervisor decides that since the one UAAV has encountered a target/threat with high threat severity, but low target priority, while the second has encountered two target/threats with higher target priority, but lower threat severity, the one UAAV ought to evade the threat, and go assist the second UAAV in identifying and possibly eliminating the high priority targets. Notice that these decisions could have been made without the supervisor UAAV, but if it is available, and has established communications with other nearby UAAVs it can make more effective decisions.

In the case, however, where there are no nearby UAAVs to communicate with, each UAAV must be able to act entirely on its own in the most effective manner possible. This situation is depicted in the southeast corner of the target/threat environment. There, we have a group of low target priority/high threat severity vehicles that are "protecting" a relatively defenseless, but high priority target (i.e., the $(6,1)$ site). Suppose that the nearby UAAV has identified two of the three in the group that is protecting the valuable target. The UAAV should then have a strategy where it tries to evade the threats and continue to seek a higher priority target, and decide in an optimal fashion when to strike targets. Hence, we see that UAAVs must be able to flexibly operate independently, but if communication is possible with nearby UAAVs, then the information should be exploited to optimize the balance between the need to identify target/threats, and to evade and destroy these.

This leads to the following principle: Wide-area search and identification of targets/threats is desirable, but it must be balanced with fuel/time constraints, threats that may dictate the need to evade and hence obtain poor identification, and the need to ultimately engage the highest priority targets with the limited resources available (e.g., for the autonomous munition problem we assume a single shot, expendable UAAV).

1.3.   Distributed guidance and control architecture

Consider N vehicles deployed in some search region $\mathcal{X}$ of known dimension. As each vehicle moves around in the search region, it obtains sensory information about the environment, which helps to reduce the uncertainty about the environment. This sensory information is typically in the form of an image, which can be processed on-line to determine the presence of a certain item or target. Alternatively, it can be in the form of a sensor coupled with automatic target recognition (ATR) software. In addition to the information received from its own sensors, each vehicle also receives information from other vehicles via a wireless communication channel. The information received from other vehicles can be in raw data form or it may be pre-processed, and it may be coming at a different rate (usually at a slower rate) than the sensor information received by the vehicle from its own sensors.

Depending on the specific mission, the global objective pursued by the team of vehicles may be different. In this paper, we focus mainly on the problem of cooperative search, where the team of vehicles seeks to follow a trajectory that would result in maximum gain in information about the environment; i.e., the objective is to minimize the uncertainty about the environment. Intuitively, each vehicle wants to follow a trajectory that leads to regions in $\mathcal{X}$ that have not been visited frequently before by the team of vehicles. The presented framework can be easily expanded to include more advanced missions such as evading threats, attacking targets, etc. In general, the team may have an overall mission that combines several of these objectives according to some desired priority.

Each vehicle has two basic control loops that are used in guidance and control, as shown in Figure 12.2. The "outer-loop" controller for vehicle $\mathcal{A}_i$ utilizes sensor information from $\mathcal{A}_i$, as well as sensor information from $\mathcal{A}_j$, $j \neq i$, to compute on-line a desired trajectory (path) to follow, which is denoted by $P_i(k)$. The sensor information utilized in the feedback loop is denoted by v_i and may include information from standard vehicle sensors (e.g. pitch, yaw, etc.) and information from on-board sensors that has been pre-processed by resident ATR software. The sensor information coming from other vehicles is represented by the vector

$$V_i = [v_1, \dots, v_{i-1}, v_{i+1}, \dots, v_N]^\mathsf{T},$$

where v_j represents the information received from vehicle $\mathcal{A}_j$. Although in the above formulation it appears that all vehicles are in range and can communicate with each other, this is not a required assumption—the same framework can be used for the case where some of the information from other vehicles is missing, or the information from different vehicles is received at different sampling rates. The desired trajectory $P_i(k)$ is generated as a digitized look-ahead path of the form

$$P_i(k) = \{p_i(k), p_i(k+1), \dots, p_i(k+q)\},$$

where $p_i(k+j)$ is the desired location of vehicle $\mathcal{A}_i$ at time $k+j$, and q is the number of look-ahead steps in the path planning procedure.

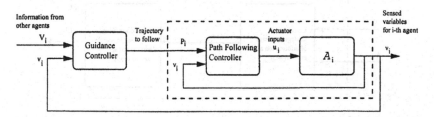

Figure 12.2. Inner- and outer-loop controllers for guidance and control of air vehicles.

The inner-loop controller uses sensed information v_i from $\mathcal{A}_i$ to generate inputs u_i to the actuators of $\mathcal{A}_i$ so that the vehicle will track the desired trajectory $P_i(k)$. We largely ignore the vehicle dynamics, and hence concentrate on the outer-loop control problem. In this way, our focus is solidly on the development of the controller for guidance, where the key is to show how resident information of vehicle $\mathcal{A}_i$ can be combined with information from other vehicles so that the team of vehicles can work together to minimize the uncertainty in the search region $\mathcal{X}$.

The design of the outer-loop control scheme is broken down into two basic functions, as shown in Figure 12.3. First, it uses the sensor information received to update its "search map", which is a representation of the environment—this will be referred to as the vehicle's *learning* function, and for convenience it will be denoted by $\mathcal{L}_i$. Based on its search map, as well as other information (such as its location and direction, the location and direction of the other vehicles, remaining fuel, etc.), the second function is to compute a desired path for the vehicle to follow—this is referred to as the vehicle's *guidance decision* function, and is denoted by $\mathcal{D}_i$. In this setting we assume that the guidance control decisions made by each vehicle are autonomous, in the sense that no vehicle tells another what to do in a hierarchical type of structure, nor is there any negotiation between vehicles. Each vehicle simply receives information about the environment from the remaining vehicles (or a subset of the remaining vehicles) and makes its decisions, which are typically based on enhancing a global goal, not only its own goal. Therefore, the presented framework can be thought of as a *passive cooperation* framework, as opposed to *active cooperation* where the vehicles may be actively coordinating their decisions and actions.

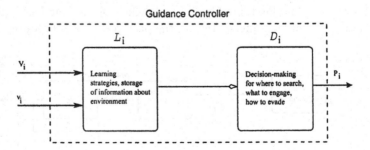

Figure 12.3. Learning and decision-making components of the outer-loop controller for trajectory generation of air vehicles.

Generally, the characteristics of the N trajectories of the group of UAAVs combine to result in the "emergence" of a set of vehicle trajectories that represent, for example:

- Cooperative search so that search resources of the group are minimized in obtaining maximum information about the environment, or

- Cooperative engagement where an attack is coordinated between UAAVs (e.g., for a particularly high priority target).

While the development of a guidance algorithm for a single UAAV is challenging, in this project it is especially important to develop learning and decision-making strategies that exploit information from other vehicles in order to realize the benefits of cooperative control for mission effectiveness enhancement.

2. Autonomous munition problem

There are several types of UAAVs at various stages of development in the US Air Force. For instance, it is envisioned that two smart Uninhabited Combat Air Vehicles (UCAVs) may be used for coordinated strike missions, and work is progressing on this topic at the Air Force Research Laboratory (AFRL). Alternatively, there is another class of UAAVs where there is a larger number of *inexpensive and expendable* autonomous munitions (AMs) with wide area search capability that could be deployed in a seek/destroy mission against critical ground mobile targets, the so-called LOCAAS problem ([17]). In our research work we use this type of UAAV problem as a benchmark for the study of design and evaluation of distributed coordination and control strategies for UAAVs. Here, we will summarize the AM problem and our current progress on developing and evaluating search and engagement strategies.

2.1. Autonomous munitions: characteristics and goals

For the AM problem we will assume that there are no threats, only mobile targets, so that we do not need to consider evasive maneuvers (i.e., so we only have a seek/destroy mission). We consider the coordination of multiple munitions ($N \geq 4$) and consider the case where there are multiple target types, with different priorities that are specified a priori. Although we assume that the targets lie in the (x, y) plane, we intend to consider terrain effects that could restrict target movement. Terrain considerations can be used to discourage munitions from searching areas where targets are not expected to be located, thus improving the overall search efficiency. We assume that each AM has a "generic sensor" (our methods will not be sensor-specific) that "illuminates" (allows it to see) only a fixed (e.g., elliptical or trapezoidal) region on the (x, y) plane (i.e., like the shaded regions in Figure 12.1). If an object is in its illumination area, the sensor will have the opportunity to recognize the object (target priority i or non-target), and if the munition declares it to be a target, provide an estimate of the (x, y) location. The generic target recognition for this project will be done using probability table look-ups, a confusion matrix, as is known to those familiar with ATR terminology. In addition to mis-identifying real targets, the munitions have the potential for declaring non-target objects to be a real target. For a declared target (real or otherwise), the minimum information we expect the munition to communicate is the target priority, (x, y) location, and whether or not the munition is committed to engaging the target.

The current state-of-the-art in munition wide area search is such that a ground mobile target free to move in any direction can increase its target location uncertainty faster (quadratic with time) than a single munition can search (linear with time). Further, each individual munition has limited sensing/computing capability with which to positively identify a target. Assuming the target is found and properly identified, each munition has a probability of kill given a hit which is less than unity. An obvious way to compensate for these factors is to use multiple munitions to help find, identify and/or strike critical targets, but this must be done in such a way that the probability of missing other targets in the area is minimized. An additional constraint on the solution is the level of communication that is possible between munitions. For example, it may be that only a low-bandwidth channel is available between munitions that are physically near each other. Currently we are also assuming range degraded communications.

The overall goal should be to increase the macro-level effectiveness (as measured by standard mission effectiveness performance metrics), while at the same time maintaining or reducing the required level of technology for each individual munition. The effectiveness of our cooperative control strategies is being compared to a baseline autonomous (non-cooperating) wide area search munition. Scenarios used as test cases are parameterized by factors such as target density, munition/target ratio, ATR performance (probability of correct target report and False Target Attack Rate, FTAR), warhead lethality, and others. The often competing objectives being addressed for each scenario are illustrated by the following cases:

1 *Maximize probability of correct target identification:* In this scenario we specify that the highest priority is to identify targets, and

that they are only destroyed if there are not further opportunities to identify more aspects of the target environment. In this case the group of AMs will likely miss the opportunity to destroy some targets, but it may be able to communicate a more accurate picture of the target environment back to a command center for a follow-on strike.

2 *Maximize the probability of high priority kills:* In this scenario the emphasis is on the destruction of the highest priority targets. For this reason, an AM finding a high priority target will immediately initiate an engagement, and a nearby munition may also decide to engage the same target. However, an AM finding a lower priority target may communicate information on the target but continue searching for a high priority target. In this case the lower priority targets will only be engaged by munitions that are running low on fuel and therefore have a low probability of finding a high priority target through continued search.

3 *Maximize overall mission effectiveness:* In this case, the set of AM's will seek to maximize a formula consisting of a weighted sum of the expected number of kills on real targets (higher priority counts more) with a negative contribution for an attack on a non-target vehicle. This case, more than the previous two, provides a very challenging problem for cooperative behavior strategies because the munition false target attack rate can result in degraded performance for the cooperative munitions as compared to the baseline autonomous munitions.

2.2. Simulation testbeds for the autonomous munitions problem

There is a simulation testbed written in FORTRAN that is currently being used at AFIT for evaluating mission effectiveness for cooperative control strategies (see [11]). The simulation (PSUB) was originally developed by Lockheed Martin Vought Systems for the LOCAAS program, and has been modified for evaluation of cooperative attack strategies. This simulation, however, is proprietary and not well suited to the cooperative search and classification problems. For these reasons, we developed an open simulation within the MATLAB environment. The capabilities of the MATLAB simulation are similar to those of PSUB, namely multiple munitions with generic target recognition capability (probability table look-ups) searching for multiple moving targets of differing priority, with user specified environmental characteristics (e.g., target densities, false target densities, etc.). Two example plots from our testbed are shown in Figure 12.4. Here we show the use of "serpentine" and "random" search for a rectangular region (the squares designate targets and triangles are vehicles).

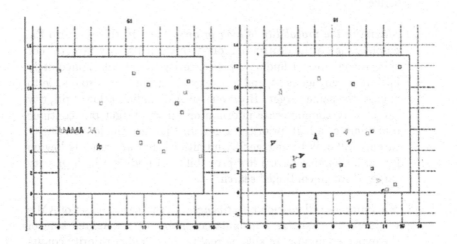

Figure 12.4.   Output from autonomous munitions simulation testbed.

The Matlab cooperative munitions simulation test bed has a friendly graphical user-interface that facilitates visualization of UAAV swarm dynamics and the effects of coordination strategy design choices. It uses object-oriented features of Matlab, has Monte-Carlo simulation abilities, and provides numerical measures of mission effectiveness. We use it in the evaluation of the effectiveness of coordination strategies, and to study trade-offs in design. A detailed description of the progress made on the development and evaluation of a "behavioral rule scheme" for the autonomous munition problem is given in the companion paper ([11]).

3.   Cooperative control via distributed learning and planning

In this section, we present a general cooperative search framework for distributed agents. This framework is used to explain how the N vehicles work together to achieve distributed learning about their environment and how distributed learning can facilitate distributed planning of UAAV activities for search and engagement missions.

3.1.   Distributed learning

Each UAAV has a three dimensional map, which we will refer to as "search map," that serves as the vehicle's knowledge base of the environment. The x and y coordinates of the map specify the location in the target environment (i.e., $(x, y) \in \mathcal{X}$), while the z coordinate specifies the certainty that the vehicle "knows" the environment at that point. The search map will be represented mathematically by an on-line approximation function as

$$z = \mathcal{S}(x, y; \theta),$$

where (x, y) is a point in the search region $\mathcal{X}$, and the output $z \in [0, 1]$ corresponds the certainty about knowing the environment at the point (x, y) in the search region. If $\mathcal{S}(x, y; \theta) = 0$ then the vehicle knows nothing (is totally uncertain) about the nature of the environment at (x, y). On the other hand, if $\mathcal{S}(x, y; \theta) = 1$ then the UAAV knows everything (or equivalently, the vehicle is totally certain) about the environment at (x, y). As the vehicle moves around in the search region it gathers new information about the environment which is incorporated into its search map. Also incorporated into its search map is the information received by communication with other vehicles. Therefore, the search map of each vehicle is continuously evolving as new information about the environment is collected and processed.

We define $\mathcal{S} : \mathcal{X} \times \Re^q \mapsto [0, 1]$ to be an on-line approximator (for example, a neural network), with a fixed structure whose input/output response is updated on-line by adapting a set of adjustable parameters, or weights, denoted by the vector $\theta \in \Re^q$. According to the standard neural network notation, (x, y) is the input to the network and z is the output of the network. The weight vector $\theta(k)$ is updated based on an on-line learning scheme, as is common, for example, in training algorithms of neural networks.

In general, the search map serves as a storage place of the knowledge that the vehicle has about the environment. While it is possible to create a simpler memory/storage scheme (without learning) that simply records the information received from the sensors, a learning scheme has some key advantages: 1) it allows generalization between points; 2) information from different types of sensors can be recorded in a common framework (on the search map) and discarded; 3) it allows greater flexibility in dealing with information received from different angles; 4) in the case of dynamic environments (for example, targets moving around), one can conveniently make adjustments to the search map to incorporate the changing environment (for example, by reducing the output value z over time using a decay factor).

The search map is formed dynamically as the vehicle moves, gathers information about the environment, and processes the information based on automatic target recognition software, or other image processing methods. This is illustrated in Figure 12.5, where we show the area scanned by a "generic" sensor on a UAAV during a sampling period $[kT, \ kT + T]$ where $T > 0$ is the sampling time. Although in different applications the shape of the scanned area maybe be different, the main idea remains the same. The received data can then be digitized and each grid point is used to adjust the search map $\mathcal{S}(x, y; \hat{\theta})$ by adapting $\hat{\theta}$.

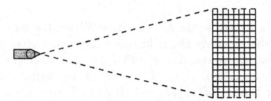

Figure 12.5. An example of a scan area for a UAAV.

In practice, the problem of minimizing the uncertainty in the search region is typically an intermediate goal. The overall objective may include, for example, finding specific targets, or avoiding certain obstacles and threats. Therefore, depending on the application being considered, the learning scheme described above for minimizing uncertainty may need to be expanded. One possible way to include a mission of searching for specific targets is to incorporate the *search map* into a more general *target search map*, which in addition to providing information about the vehicle's knowledge of the environment, it also contains information about the presence (or not) of targets. This can be achieved by allowing the output z of the on-line approximator $\mathcal{S}$ to take values in the region $z \in [-1, \ 1]$, where:

- $z = \mathcal{S}(x, y; \theta) = 1$ represents high certainty that a target is present at (x, y);

- $z = \mathcal{S}(x, y; \theta) = -1$ represents high certainty that a target is not present at (x, y);

- $z = \mathcal{S}(x, y; \theta) = 0$ represents total uncertainty whether a target is present at (x, y).

This representation contains additional information that the vehicle can utilize in making guidance and path planning decisions. Furthermore, the learning framework can be extended to a multi-dimensional framework, where the output z of the on-line approximator is a vector of dimension greater than one. For example, one could use the first output to represent the presence/absence of a target (as above), and the second output to represent the priority of the target.

In this general framework, the tuning of the search map can be viewed as "learning" the environment. Mathematically, $\mathcal{S}$ tries to approximate an unknown function $\mathcal{S}^*(x, y, k)$, where for each (x, y), the function $\mathcal{S}^*$ characterizes the presence (or not) of a target; the time variation indicated by the time step k is due to (possible) changes in the environment (such as having moving targets). Hence, the learning problem is defined in using sensor information from vehicle A_i and information coming from other vehicles A_j, $j \neq i$ at each sampled time k, to adjust the weights $\hat{\theta}(k)$ such that

$$\left\| \mathcal{S}(x, y; \hat{\theta}(k)) - \mathcal{S}^*(x, y, k) \right\|_{(x,y) \in \mathcal{X}}$$

is minimized.

Due to the nature of the learning problem, it is convenient to use spatially localized approximation models so that learning in one region of the search space does not cause any "unlearning" at a different region ([36]). The dimension of the input space (x, y) is two, and therefore there are no problems related to the "curse of dimensionality" that are usually associated with spatially localized networks. In general, the learning problem in this application is straightforward, and the use of simple approximation functions and learning schemes is sufficient; e.g., the use of piecewise constant maps or radial basis function networks, with distributed gradient methods to adjust the parameters, provides sufficient learning capability. However, complexity issues do arise and are crucial since the distributed nature of the architecture imposes limits not only on the amount of memory and computations needed to store and update the maps but also in the transmission of information from one vehicle to another.

At the time of deployment, it is assumed that each vehicle has a copy of an initial search map estimate, which reflects the current knowledge about the environment $\mathcal{X}$. In the special case that no a priori information is available, then each point on the search map is initialized as "completely uncertain." In general, each vehicle is initialized with the same search map. However, in some applications it may be useful to have UAAVs be "specialized" to search in certain regions, in which case the search environment for each UAAV, as well as the initial search map, may be different.

3.2.   Cooperative path planning

One of the key objectives of each vehicle is to on-line select a suitable path in the search environment $\mathcal{X}$. To be consistent with the motion dynamics of air vehicles, it is assumed that each vehicle has limited maneuverability, which is represented by a maximum angle θ_m that the vehicle can turn from its current direction. For simplicity we assume that all vehicles move at a constant velocity μ (this assumption can be easily relaxed).

3.2.1   Plan Generation.   To describe the movement path of vehicle $\mathcal{A}_i$ between samples, we define the *movement sampling time* T_m as the time interval in the movement of the vehicle. In this framework, we let $p_i(k)$ be the position (in terms of (x, y) coordinates) of i-th vehicle at time $t = kT_m$, with the vehicle following a straight line in moving from $p_i(k)$ to its new position $p_i(k + 1)$. Since the velocity μ of the UAAV is constant, the new position $p_i(k + 1)$ is at a distance μT_m from $p_i(k)$, and based on the maneuverability constraint, it is within an angle $\pm\theta_m$ from the current direction, as shown in Figure 12.6. To formulate the optimization problem as an integer programming problem, we discretize the arc of possible positions for $p_i(k + 1)$ into m points, denoted by the set

$$\overline{\mathcal{P}}_i(k+1) = \left\{ \bar{p}_i^1(k+1),\ \bar{p}_i^2(k+1),\ \ldots\ \bar{p}_i^j(k+1),\ \ldots \bar{p}_i^m(k+1) \right\}.$$

Therefore, the next new position for the i-th vehicle belongs to one of the elements of the above set; i.e., $p_i(k + 1) \in \overline{\mathcal{P}}_i(k + 1)$.

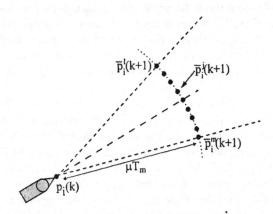

Figure 12.6.   Selection of the next point in the path of the vehicle.

The UAAV selects a path by choosing among a possible set of future position points. In our formulation we allow for a recursive q-step ahead planning, which can be described as follows:

- When vehicle $\mathcal{A}_i$ is at position $p_i(k)$ at time k, it has already decided the next q positions: $p_i(k+1)$, $p_i(k+2)$, ..., $p_i(k+q)$.

- While the vehicle is moving from $p_i(k)$ to $p_i(k+1)$ it selects the position $p_i(k+q+1)$, which it will visit at time $t = k+q+1$.

To get the recursion started, the first q positions, $p_i(1)$, $p_i(2)$, ..., $p_i(q)$ for each vehicle need to be selected a priori. Clearly, $q = 1$ corresponds to the special case of no planning ahead. The main advantage of a planning ahead algorithm is that it creates a buffer for path planning. From a practical perspective this can be quite useful since air vehicle require (at least) some trajectory planning. On the other hand, if the integer q is too large then, based on the recursive procedure, the position $p_i(k)$ was selected q samples earlier at time $k - q$; hence the decision may be outdated, in the sense that it may have been an optimal decision at time $k - q$, but based on the new information received since then, it may not be the best decision anymore. The recursive q-step ahead planning procedure is illustrated in Figure 12.7 for the case where $q = 6$.

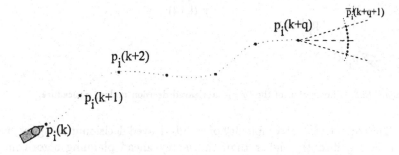

Figure 12.7. Illustration of the recursive q-step ahead planning algorithm.

If practical considerations (such as motion dynamics of the vehicle and computational demands for path selection) require a relatively large value for q then the problem of "outdated" decision making can be ameliorated by an *interleaved* type of scheme. We define a (q, r)-interleaved decision making scheme as follows:

- When vehicle $\mathcal{A}_i$ is at position $p_i(k)$ at time k, it has already decided the next q positions: $p_i(k+1)$, $p_i(k+2)$, ... $p_i(k+q)$.

- While the vehicle is moving from $p_i(k)$ to $p_i(k+1)$ it re-calculates the last r points of the path based on the current data and also selects another new position; i.e., it selects the points $p_i(k+q-r+1)$, $p_i(k+q-r+2)$, ..., $p_i(k+q)$, $p_i(k+q+1)$.

The term "interleaved" is used to express the fact that decisions are re-calculated over time, as the vehicle moves, to incorporate new information that may have been received about the environment. According to this formulation, a (q, r)-interleaved decision scheme requires the selection of $r+1$ points for path planning at each sample T_m. The special case of $(q, 0)$-interleaved scheme (actually, strictly speaking it is a non-interleaved scheme) corresponds to the recursive q-step ahead planning scheme described earlier. Similar to the recursive q-step ahead planning scheme, at the beginning, the first q positions for each vehicle need to be selected a priori. The interleaved path planning procedure is illustrated in Figure 12.8 for the case where $q = 6$ and $r = 2$.

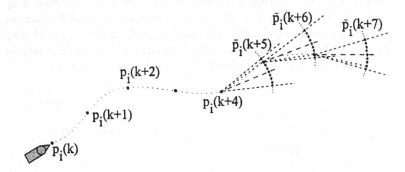

Figure 12.8. Illustration of the (q, r)-interleaved decision making procedure.

The computational complexity of an interleaved decision making scheme can be significantly higher than the q-step ahead planning algorithm. Specifically, with the q-step ahead planning algorithm, each vehicle has to select one position among m possible candidates. With the (q, r)-interleaved algorithm, each vehicle has to select $r+1$ positions among a combination of m^{r+1} candidates. Therefore, the computational complexity increases exponentially with the value of the interleaved variable r. This is shown in Figure 12.8 where $m = 9$, $r = 2$; therefore at each sample time the vehicle needs to select among $9^3 = 243$ possible paths in order to compute the three positions $p_i(5)$, $p_i(6)$ and $p_i(7)$. The figure shows a path of points generated by the guidance (outer-loop) controller, and then shows a tree of possible directions that the vehicle can take.

3.2.2   Plan Selection.   Given the current information available via the search map, and the location/direction of the team of vehicles (and possibly other useful information, such as fuel remaining, etc.), each vehicle uses a multi-objective cost function J to select and update its search path. At decision sampling time T_d, the vehicle evaluates the cost function associated with each path and selects the optimal path. The decision sampling time T_d is typically equal to the movement sampling time T_m. The approach can be thought of as an "adaptive model predictive control" approach where we learn the model that we use to predict ahead in time, and we use on-line optimization in the formation of that model, and in evaluating the candidate paths to move the vehicle along.

A key issue in the performance of the cooperative search approach is the selection of the multi-objective cost function associated with each possible path. Our approach is quite flexible in that it allows the characterization of various mission-level objectives, and trade-offs between these. In general, the cost function comprises of a number of sub-goals, which are sometimes competing. Therefore the cost criterion J can be written as:

$$J = \omega_1 J_1 + \omega_2 J_2 + \cdots + \omega_s J_s,$$

where J_i represents the cost criterion associated with the i-th subgoal, and ω_i is the corresponding weight. The weights are normalized such that $0 \leq \omega_i \leq 1$ and the sum of all the weights is equal to one; i.e., $\sum_{i=1}^{s} \omega_i = 1$. Priorities to specific sub-goals are achieved by adjusting the values of weights ω_i associated with each sub-goal.

The following is a list (not exhaustive) of possible sub-goals that a search vehicle may include in its cost criterion. Corresponding to each sub-goal is a cost-criterion component that can be readily designed. For a more clear characterization, these sub-goals are categorized according to three mission objectives: Search (S), Cooperation (C), and Engagement (E). In addition to sub-goals that belong purely to one of these classes, there are some that are a combination of two or more missions. For example, SE1 (see below) corresponds to a search and engage mission.

S1 *Follow the path where there is maximum uncertainty in the search map.* This cost criterion simply considers the uncertainty reduction associated with the sweep region between the current position $p_i(k)$ and each of the possible candidate positions $\bar{p}_i^j(k+1)$ for the next sampling time (see the rectangular regions between $p_i(k)$ and $\bar{p}_i^j(k+1)$ in Figure 12.9). The cost criterion can be derived by computing a measure of uncertainty in the path between $p_i(k)$ and each candidate future position $\bar{p}_i^j(k+1)$.

S2 *Follow the path that leads to the region with the maximum uncertainty (on the average) in the search map.* The first cost cost criterion (S1) pushes the vehicle towards the path with the maximum uncertainty. However, this may not be the best path, over a longer period of time, if it leads to a region where the average uncertainty is low. Therefore, it's important for the search vehicle to seek not only the instantaneous minimizing path, but also a path that will cause the vehicle to visit (in the future) regions with large uncertainty. The cost criterion can be derived by computing the average uncertainty of a triangular type of region associated with the heading direction of the vehicle (see the triangular regions ahead of $\bar{p}_i^j(k+1)$ in Figure 12.9).

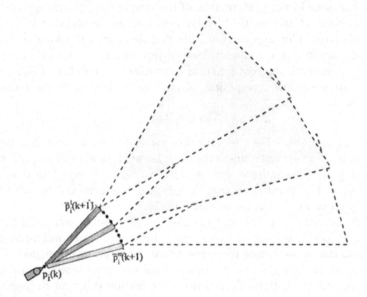

Figure 12.9. Illustration of the regions that are used in the cost function for finding the optimal search path.

C1 *Follow the path where there is the minimum overlap with other vehicles.* Since the vehicles are in frequent communication with each other, they are able to share their new information about the search region, thereby the search map of each individual vehicle is similar to the search maps of the other vehicles. Consequently, it is natural that they may select the same search path as other vehicles (especially since in general they will be utilizing the same search algorithm). This will be more pronounced if two vehicles happen to be close to each other. However, in order to minimize the global uncertainty associated with the emergent knowledge of all vehicles, it is crucial that there is minimum overlap in their search efforts. This can be achieved by including a cost function component that penalizes vehicles being close to each other and heading in the same direction. This component of the cost function can be derived based on the relative locations and heading direction (angle) between pairs of vehicles.

SE1 *Follow the path that maximizes coverage of the highest priority targets.* In mission applications where the vehicles have a target search map with priorities assigned to detected targets, it is possible to combine the search of new targets with coverage of discovered targets by including a cost component that steers the vehicle towards covering high priority targets. Therefore, this leads to a coordinated search where both coverage and priorities are objectives.

E1 *Follow the path toward highest priority targets with most certainty if fuel is low.* In some applications, the energy of the vehicle is a key limiting factor. In such cases it is important to monitor the remaining fuel and possibly switch goals if the fuel becomes too low. For example, in search-and-engage operations, the vehicle may decide to abort search objectives and head towards engaging high priority targets if the remaining fuel is low.

EC1 *Follow the path toward targets where there will be minimum overlap with other vehicles.* Cooperation between vehicles is a key issue not only in search patterns but also—and even more so—in engagement patterns. If an vehicle decides to engage a target, there needs to be some cooperation such that no other vehicle tries to go after the same target; i.e., a coordinated dispersed engagement is desirable.

The above list of sub-goals and their corresponding cost criteria provide a flavor of the type of issues associated with the construction of the overall cost function for a general mission. In addition to incorporating the desired sub-goals into the cost criterion (i.e., maximize benefit), it is also possible to include cost components that reduce undesirable sub-goals (minimize cost). For example, in order to generate a smooth trajectory for a UAV such that it avoids—as much as possible—the loss of sensing capabilities during turns, it may be desirable to assign an extra cost for possible future positions on the periphery (large angles) of the set $\overline{\mathcal{P}}_i$.

In the next subsection, we consider some simulation studies that are based on a cost function consisting of the first three sub-goals. Therefore the main goal is to search in a cooperative framework.

3.3.   Cooperative search and learning results

The proposed cooperative search and learning framework has been tested in several simple simulated studies. Three of these simulation studies are presented in this subsection. In the first simulation experiment there are two vehicles, while in the second simulation we will be using a team of five vehicles. In both of these simulation studies we are using the recursive q-step ahead planning algorithm with $q = 3$. In the final simulation experiment we compare the q-step ahead planning algorithm with the interleaved scheme, and also examine the effect of a dynamic environment, where the location of targets maybe changing over time.

The results for the case of two vehicles are shown in Figure 12.10. The upper-left plot shows a standard search pattern for the first 500 time samples, while the upper-right plot shows the corresponding result for a random search, which is subject to the maneuverability constraints. The standard search pattern utilized here is based on the so-called zamboni coverage pattern ([1]). The lower-left plot shows the result of the cooperative search method based on the recursive q-step ahead planning algorithm. The search region is a 200 by 200 area, and we consider two search vehicles which start at the location indicated by the triangles. It is assumed that there is some a priori information about the search region: the green (light) polygons indicate complete certainty about the environment (for example, these can represent regions where it is known for sure—due to the terrain—that there are no targets); the blue (dark) polygons represent partial certainty about the environment. The remaining search region is assumed to be completely uncertain. For simplicity, in this simulation the only mission is for the vehicles to work cooperatively in reducing the uncertainty in the environment (by searching in highly uncertain areas).

The search map used in this simulation study is based on piecewise constant basis functions, and the learning algorithm is a simple update algorithm of the form $\hat{\theta}(k+1) = 0.5\hat{\theta}(k)+0.5$, where the first encounter of a search block results in the maximum reduction in uncertainty. Further encounters result in reduced benefit. For example, if a block on the search map starts from certainty value of zero (completely uncertain) then after four visits from (possibly different) vehicles, the certainty value changes to $0 \mapsto 0.5 \mapsto 0.75 \mapsto 0.875 \mapsto 0.9375$. The percentage of uncertainty

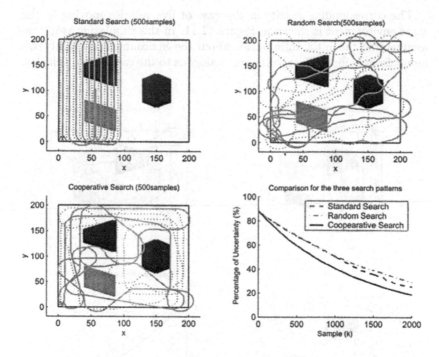

Figure 12.10. Comparison of the cooperative search pattern with a "standard" search pattern and a random search pattern for the case of two moving vehicles. The upper-left plot shows a standard search pattern for the first 500 time samples; the upper-right plot shows the corresponding search pattern in the case of a random search, subject to some bounds to restrict the vehicle from deserting the search region; The lower-left plot shows the cooperative search pattern based on the recursive q-step ahead planning algorithm; the lower-right plot shows a comparison of the performance of the three search patterns in terms of reducing uncertainty in the environment.

is defined as the distance of the certainty value from one. In the above example, after four encounters the block will have 6.25% percentage of uncertainty. The cooperative search algorithm has no pre-set search pattern. As seen from Figure 12.10, each vehicle adapts its search path on-line based on current information from its search results, as well as from search results of the other vehicles.

To compare the performance of the three search patterns, the lower-right plot of Figure 12.10 shows the percentage of uncertainty with time for the standard search pattern, the random search pattern and the cooperative search pattern described above. The ability of the cooperative search algorithm to make path planning decisions on-line results in a faster rate of uncertainty reduction.

The corresponding results in the case of five vehicles moving in the same environment is shown in Figure 12.11. In this simulation study we assume that the initial information about the environment is slightly different from before. The results are analogous to the case of two vehicles.

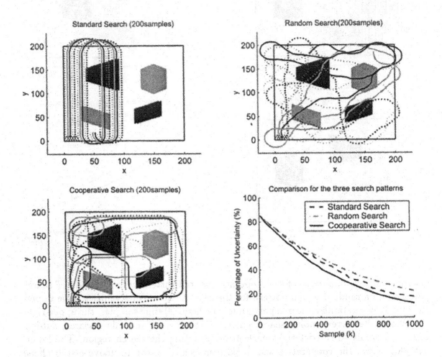

Figure 12.11. Comparison of the cooperative search pattern with a "standard" search pattern and a random search pattern for the case of five moving vehicles. The upper-left plot shows a standard search pattern for the first 500 time samples; the upper-right plot shows the corresponding search pattern in the case of a random search, subject to some bounds to restrict the vehicle from deserting the search region; The lower-left plot shows the cooperative search pattern based on the recursive q-step ahead planning algorithm; the lower-right plot shows a comparison of the performance of the three search patterns in terms of reducing uncertainty in the environment.

In these simulation studies, we assume that the sampling time $T_m = 1$ corresponds to the rate at which each vehicle receives information from its own sensors, updates its search map and makes path planning decisions. Information from other vehicles is received at a slower rate. Specifically, we assume that the communication sampling time T_c between vehicles is five times the movement sampling time; i.e., $T_c = 5T_m$. For fairness in comparison, it is assumed that for the standard and random search patterns the vehicles exchange information and update their search maps in the same way as in the cooperative search pattern, but they do not use the received information to make on-line decisions on where to go.

In the first two simulations it was assumed that learning of the environment was monotonically increasing with time; in other words, as the team of UAAVs moved around the environment, they were able to enhance their knowledge, leading to decreasing levels of uncertainty. In the case of a dynamic environment the targets may be moving around, and therefore the represented level of uncertainty increases with time. In the framework followed in this paper, reduction in the certainty levels due to a dynamic environment can be implemented by using a decay value $d < 1$ that multiplies the search map; i.e., $z = \mathcal{S}(x, y, \hat{\theta})d$. Therefore, in the case of a dynamic environment, learning (which reduces uncertainty) competes with the increase in uncertainty due to a changing environment. In the third simulation experiment, shown in Fugure 12.12, we consider the same environment as in Figure12.11 with five UAAVs, but introduce a decay to represent a dynamic environment. The simulation shown is for three different levels of decay: (i) $d = 0.98$; (ii) $d = 0.995$; and (iii) $d = 1$ (no decay). This decay factor is applied every 5 steps. For different levels of decay, we compare the non-interleaved planning algorithm $r = 0$ with the interleaved search planning algorithm $r = 1$. As expected, the rate of reduction in percentage of uncertainty decreases as the decay rate d decreases. The plot also shows that, overall, the interleaved search planning algorithm performs slightly better than the corresponding non-interleaved.

It is noted that in these simulations the path planning of the cooperative search algorithm is rather limited since at every sampled time each vehicle is allowed to either go straight, left, or right (the search direction is discretized into only three possible points; i.e., $m = 3$). As the complexity of the cooperative search algorithm is increased and the design parameters (such as the weights associated with the multi-objective cost function) are fine-tuned or optimized, it is anticipated that the search performance can be further enhanced.

3.4. Related research work on search methods

Various types of search problems occur in a number of military and civilian applications, such as search-and-rescue operations in open-sea or sparsely populated areas, search missions for previously spotted enemy targets, seek-destroy missions for land mines, and search for mineral deposits. A number of approaches have been proposed for addressing such search problems. These include, among other, optimal search theory

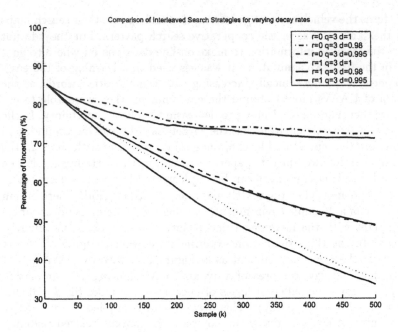

Figure 12.12. Comparison of interleaved search strategies for varying decay rates.

([30, 19]), exhaustive geographic search ([27]), obstacle avoidance ([4, 26]) and derivative-free optimization methods ([6]).

Search theory deals with the problem of distribution of search effort in a way that maximizes the probability of finding the object of interest. Typically, it is assumed that some prior knowledge about the target distribution is available, as well as the "payoff" function that relates the time spent searching to the probability of actually finding the target, given that the target is indeed in a specific cell ([30, 19]). Search theory was initially formed during World War II with the work of Koopmam and his colleagues at the Anti-Submarine Warfare Operations Research Group (ASWORG). Later on, the principles of search theory were applied successfully in a number of applications, including the search for and rescue of a lost party in a mountain or a missing boat on the ocean, the surveillance of frontiers or territorial seas, the search for mineral deposits, medical diagnosis, and the search for a malfunction in an industrial process. Detailed reviews of the current status of search theory are given by ([31, 25, 2]).

The optimal search problem can be naturally divided according to two criteria that depend on the target's behavior. The first division depends on whether the target is evading or not; that is, whether there is a two-sided optimization by both the searcher and the target, or whether the target's behavior is independent of the searcher's action. The second division deals with whether the target is stationary or moving. The two divisions and their combinations form four different categories. A great deal of progress in solving stationary target problems in the optimal search framework has been made, and solutions have been derived for most of the standard cases ([30]). For the moving target problem, the emphasis in search theory has shifted from mathematical and analytical solutions to algorithmic solutions ([2]). A typical type of search problem, called the path constrain search problem (PCSP), that takes into account the movement of the searcher, was investigated by several researchers ([10, 29, 15, 14]). Because of the NP-complete nature of this problem, most authors proposed a number of heuristic approaches that result in approximately optimal solutions. The two-sided search problem can be treated as a game problem for both the searcher and target strategies. This has been the topic of a number of research works ([7, 16, 35]). So far, search theory has paid little attention to the problem of having a team of cooperating searchers. A number of heuristic methods for solving this problem have been proposed by ([8]).

The Exhaustive Geographic Search problem deals with developing a complete map of all phenomena of interest within a defined geographic area, subject to the usual engineering constraints of efficiency, robustness and accuracy ([27]). This problem received much attention recently, and algorithms have been developed that are cost-effective and practical. Application examples of Exhaustive Geographic Search include mapping mine fields, extraterrestrial and under-sea exploration, exploring volcanoes, locating chemical and biological weapons and locating explosive devices ([27, 12, 13, 5]).

The obstacle avoidance literature deals with computing optimal paths given some kind of obstacle map. The intent is to construct a physically realizable path that connects the initial point to the destination in a way that minimizes some energy function while avoiding all the obstacles along the route ([4, 26]). Obstacle avoidance is normally closely geared to the methods used to sense the obstacles, as time-to-react is of the essence. The efficiency of obstacle avoidance systems is largely limited by the reliability of the sensors used. A popular way to solve the obstacle avoidance problem is the potential field technique ([18]). According to the potential field method, the potential gradient that the robot follows is made up of two components: the repulsive effect of the obstacles and the attractive effect of the goal position. Although it is straightforward to use potential field techniques for obstacle avoidance, there are still several difficulties in using this method in practical vehicle planning.

Derivative-Free Optimization methods deal with the problem of minimizing a nonlinear objective function of several variables when the derivatives of the objective function are not available ([6]). The interest and motivation for examining possible algorithmic solutions to this problem is the high demand from practitioners for such tools. The derivatives of objective function are usually not available either because the objective function results from some physical, chemical or economical measurements, or, more commonly, because it is the result of a possibly very large and complex computer simulation. The occurrence of problems of this nature appear to be surprisingly frequent in the industrial setting. There are several conventional deterministic and stochastic approaches to perform optimization without the use of analytical gradient information or measures of the gradient. These include, for example, the pattern and coordinate search ([33, 20]), the Nelder and Mead Simplex Method ([22]), the Parallel Direct Search Algorithm ([9]), and the Multi-directional Search Method ([32]). In one way or another most of derivative free optimization methods use measurements of the cost function and form approximations to the gradient to decide which direction to move. ([24]) provides some ideas on how to extend non-gradient methods to team foraging.

4.   Stable vehicular swarms

Groups of communicating vehicles, such as automobiles (e.g., in "platoons" in automated highway systems), robots, underwater vehicles, or aircraft formations have been studied for some time. In the current project we are conducting an analytical study where we are mathematically modeling "vehicular swarms" (groups of vehicles that behave as a single entity) where there are communication delays, and deriving conditions under which the swarms are stable (cohesive). The central focus of our work is not "swarming" per se, but the factors that drive a group of UAAVs to decide to aggregate (e.g., work together to search an area, or make a coordinated attack) or disintegrate (e.g., spread out to search, or scatter to evade a threat). Such factors include characteristics of the vehicle-to-vehicle communications, target/threat priorities and densities, and mission goals.

4.1. Stable asynchronous vehicular swarms with communication delays

To briefly summarize our current work we first explain how stability relates to swarm cohesion, aggregation, and disintegration. Next, we overview the types of stability conditions we obtain for swarms, and show the output of a simulation testbed we have developed for studying swarm dynamics.

Stability, Cohesion, Aggregation, and Disintegration:. To achieve coordinated actions, a group of autonomous vehicles requires vehicle-to-vehicle communications. When should a swarm of UAAVs aggregate and when should it disintegrate? Some examples include:

- It could enhance mission effectiveness to aggregate for a coordinated attack on high priority target, or to protect one another from certain types of threats (e.g., vehicles on the edges of the swarms could protect ones in the middle). Or, it may be beneficial to work together to search an area.

- In some situations mission effectiveness could be enhanced if the swarm disintegrates (breaks up, disperses) for good search coverage (i.e., so search areas do not overlap) when there are many similar stationary targets with a uniform density over the search area. At other times it may be beneficial for a swarm to disintegrate in order to evade certain types of threats (e.g., a threat that has the ability to kill a whole group of UAAVs, in an especially effective way if they are all grouped together).

- Local UAAV goals and overall mission goals, which may compete with each other, affect the dynamics of vehicle swarm aggregation and disintegration. For instance, if the mission goals dictate that only the highest priority targets should be engaged, then for some coordination strategies it may be that there is a higher likelihood of aggregation near high priority targets for coordinated attacks on them.

A problem of critical importance is then to determine how characteristics of vehicle-to-vehicle communications affect the ability of a group of UAAVs to aggregate and disintegrate. Can aggregation/disintegration be achieved when the sensor ranges for each UAAV is limited? How does the sensor range impact formation of multiple swarms, and the ability to group vehicles for a coordinated attack? What are the effects of communication delays on the dynamic behavior of a swarm, and its effectiveness?

Moreover, it is clear that characteristics of the target/threat environment affect swarm aggregation/disintegration. Can we characterize a relationship between target density and priority and when it is good to swarm, or how this will drive a group of UAAVs to swarm? What threat patterns should drive the disintegration of a swarm of UAAVs? Finally, mission goals affect the overall dynamics of the vehicular swarm. How can we make sure that mission goals are properly specified so that detrimental aggregation/disintegration events do not occur (e.g., aggregation near a threat so that an entire group of UAAVs could be easily eliminated), and hence mission effectives is maintained.

In summary, stability analysis focuses on the underlying mechanisms that affect the dynamics of the coordinated behavior of the *group* of UAAVs. It is of fundamental importance to understand how to exploit the benefits of coordinated search and engagement for a group of UAAVs.

Stability of Swarms Modeled as Distributed Discrete Event Systems:.   Here, our approach to begin to study such problems is to introduce a discrete event system (DES) framework based on the approach in ([23]), characterize swarm cohesion characteristics via stability properties, then use relatively conventional stability analysis methods (e.g., Lyapunov methods) to provide conditions under which various cohesion properties can be achieved. The key theoretical difficulty is how to perform stability analysis for distributed DES with communication delays; however, the approaches in ([23, 34]) provide a basis to start, and this is the approach that we take.

A two-dimensional (2-D) swarm is formed by putting many single swarm members together on the (x_1, x_2)-plane. Assume that we have a two-dimensional asynchronous N-member swarm, where $N > 1$. Let $x^i(t)$ denote the position vector of swarm member i at time t. Let T^i be the set of time indices at which the ith vehicle moves. We have $x^i(t) = [x_1^i(t), x_2^i(t)]^\top \in \Re^2$, $i = 1, 2, ..., N$, where $x_1^i(t)$ and $x_2^i(t)$ are member i's horizontal and vertical position coordinates respectively. The single swarm member model used here to discuss some of the basic ideas of our work includes only the simplest of dynamics (point-mass), neighbor position sensors, and proximity sensors. For a swarm member, we assume all other members which can be sensed by its neighbor position sensors are its "neighbors," and its sensed position information about its neighbors may be subjected to random but bounded delays. Define $\tau_j^i(t)$ to be the last time at which vehicle i received position information from vehicle j, where $i, j = 1, 2, ..., N$, $j \neq i$. Assume the proximity sensor of the swarm members has a circular-shaped sensing range with a radius $\varepsilon > 0$, which can instantaneously indicate the positions of any other members inside this range. Swarm members like to be close to each other, but not too close. Assume each member has a "private" area, which is a circular-shaped area with a radius $d > 0$ around each member, where d is the desired comfortable distance between two adjacent swarm neighbors, which is known by every swarm member. For some $\gamma > 0$, assume $[d, d + \gamma]$ is a "comfortable distance neighborhood" relative to $x^i(t)$ and $x^j(t)$ for all i and j, where γ is the comfortable distance neighborhood size. In other words, swarm members do not want other members to enter its private area and they prefer their neighbors are at a comfortable distance neighborhood $[d, d + \gamma]$ from them. Here, for the sake of illustration we choose $\gamma = d = \varepsilon/2$. We assume that $|x| = \sqrt{x^T x}$ and $|x^i(0) - x^j(0)| > d$, for $i, j = 1, 2, ..., N, i \neq j$ initially. We make certain assumptions about the times that all the vehicles will update their position and bounds on the length of the delays for obtaining neighbor's position information (similar to the "partial asynchronism" assumption in ([34])).

Each swarm member i, $i = 1, 2, ..., N$, remains stationary at the beginning until some $t' \in T^i$, where it has sensed all its $N - 1$ neighbors' position information and it then calculates the center position of them as its goal position. Assume the goal position $x_c^i(t)$ of member i at time $t \in T^i$ is defined as

$$x_c^i(t) = \frac{1}{N-1} \sum_{j=1, j \neq i}^{N} x^i(\tau_j^i(t)), \forall t \in T^i, i = 1, 2, ..., N \qquad (1)$$

Member i tries to approach this goal position since it wants to be adjacent to all its neighbors. Note that $x_c^i(t)$ may not be the real center of its neighbors since the swarm member's sensed information may include random delays. Assume swarm member i will move toward $x_c^i(t)$ with a step size d, which is the radius of its private area if it finds the distance between them is larger than or equal to d. It will move to $x_c^i(t)$ with one step if it finds the distance is less than d. It will remain stationary if it is already at $x_c^i(t)$. During movements it does not want any other member to enter into its private area because swarm members like to be close to each other, but not so close as to enter others' private area so that collisions can occur. Hence, assume that before swarm member i moves to a new position it will detect if there is any member inside the private area around this new position via its proximity sensors (note that we choose $\varepsilon = 2d$ so that the sensing range of proximity sensors is large enough to detect the new private area). It moves to this new position only when no member is found in the new private area. Otherwise, it remains stationary.

From the above moving rules, we assume the step vector $\phi^i(t)$ of member i at time $t \in T^i$ is such that

$$\phi^i(t) = \min\{d, |x_c^i(t) - x^i(t)|\} \left[\frac{x_c^i(t) - x^i(t)}{|x_c^i(t) - x^i(t)|} \right],$$

$$\forall t \in T^i, i = 1, 2, ..., N \tag{2}$$

Here, the term in brackets is a unit vector which represents the moving direction, where $x_c^i(t)$ is the goal position of member i at time t defined in Equation (1), and the item in front of the brackets is a step size scalar, where "min" is used to model the rules of how to choose the step size. The step size is equal to d if $|x_c^i(t) - x^i(t)| \geq d$, and is equal to $|x_c^i(t) - x^i(t)|$ if $|x_c^i(t) - x^i(t)| < d$.

A mathematical model for the above asynchronous mobile swarm is given by

$$x^i(t+1) = x^i(t) + \phi^i(t) \prod_{j=1, j \neq i}^{N} u(|x^i(t) + \phi^i(t) - x^j(t)| - d),$$

$$\forall t \in T^i, i = 1, 2, ..., N \tag{3}$$

where "u" is the step function, which is equal to one when the function variable is larger than or equal to zero, and is equal to zero when the function variable is less than zero. Clearly, swarm member i will remain stationary if it detects any neighbor in its new private area, and will move to a new position $x^i(t) + \phi^i(t)$ if no neighbor is found in its new private area.

4.2.   Stability analysis and simulation results

We have derived conditions involving bounds on communication delays, and characteristics of the asynchronous behavior of the communications, that are sufficient to guarantee the ability of a 1-D asynchronous vehicular swarm to aggregate. Our approach is based on defining a Lyapunov-like function for the inter-vehicle distance for two swarm members, showing properties for this case, and then using an induction method to extend it to the multiple-swarm member case. We show that in some cases intervehicle distances converge to a desired constant (i.e., asymptotic stability), while if communications are degraded such distances only converge to a pre-specified neighborhood (i.e., uniform ultiimate boundedness). We are currently extending these results to the two (and M) dimensional cases and investigating how the design of the coordination strategies affect stability conditions.

In addition, we have developed a simulation testbed for gaining insights into the dynamics of vehicular swarms. Below, in Figure 12.13 we show a sequence of plots to illustrate swarm dynamics for a 2-D swarm. Here, we show one lead vehicle (the diamond) moving across a plane, and there are several other vehicles that swarm together and follow the lead vehicle. Such simulation results show how: (i) aggregation is achieved, (ii) the dynamics of cohesion (e.g., when moving as a group do the communication delays cause unacceptable oscillations in inter-vehicle distance), and (ii) characteristics of motion of a mobile swarm (e.g., how motion in one direction "stretches" the swarm along the velocity vector of the group, how coordinated motion can slow UAAV velocities since adjustments made to maintain cohesion result in movements to avoid collisions that could be opposite to the velocity vector of the swarm).

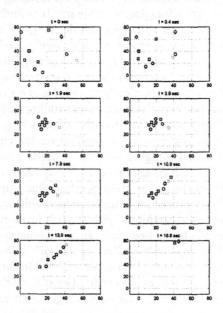

Figure 12.13. Mobile vehicular swarm (plots show a sequence of eight positions of the vehicles over time).

5. Biomimicry of foraging for cooperative control

Animal species of all sorts have been involved in daily "warfare" for millions of years when they seek food (prey) and try to avoid noxious substances or predators (i.e., when they "forage"). What can we learn from the evolved behaviors, especially in the case of "social foraging" where groups of animals work together to capture more prey (in our domain, kill targets) and at the same time avoid predators (in our domain, evade threats)? In this section we will briefly overview some initial work done to understand the relevance of social foraging behaviors of animals to the cooperative control problem for UAAVs. For more details see ([24]).

5.1.   Optimal decision-making in foraging

Animals search for and obtain nutrients in a way that maximizes $\frac{E}{T}$ where E is energy obtained, and T is time spent foraging (or, they maximize long-term average rate of energy intake) ([28]). Evolution optimizes foraging strategies since animals that have poor foraging performance do not survive. Generally, a foraging strategy involves finding a "patch" of food (e.g., group of bushes with berries), deciding whether to enter it and search for food (do you expect a better one?), and when to leave the patch. There are predators and risks, energy required for travel, and physiological constraints (sensing, memory, cognitive capabilities). Foraging scenarios can be mathematically modeled and optimal policies can be found using, for instance, dynamic programming ([28]).

Some animals forage as individuals, others forage as groups, and some will choose which approach to use depending on the current situation. While to perform social foraging an animal needs communication capabilities, it can gain advantages in that it can essentially exploit the sensing capabilities of the group, the group can "gang-up" on large prey, individuals can obtain protection from predators while in a group, and in a certain sense the group can forage with a type of collective intelligence. Social foragers include birds, bees, fish, ants, wildebeasts, and primates. Note that there is a type of "cognitive spectrum" where some foragers have little cognitive capability, and other higher life forms have significant capabilities (e.g., compare the capabilities of a single ant with those of a human). Generally, endowing each forager with more capabilities can help them succeed in foraging, both as an individual and as a group.

5.2.   E. coli bacterial foraging strategies: useful for cooperative control of UAAVs?

Consider the foraging behavior ("chemotaxis") of *E. coli*, which is a common type of bacteria that takes the following foraging actions (these are "behavioral rules"):

1 If in neutral medium alternate tumbles and runs ⇒ Search

2 If swimming up nutrient gradient (or out of noxious substances) swim longer (climb up nutrient gradient or down noxious gradient) ⇒ Seek increasingly favorable environments

3 If swimming down nutrient gradient (or up noxious substance gradient), then search ⇒ Avoid unfavorable environments

What is the resulting emergent pattern of behavior for a whole group of *E. coli* bacteria? Generally, as a group they will try to find food and avoid harmful phenomena, and when viewed under a microscope you will get a sense that a type of intelligent behavior has emerged since they will seem to intentionally move as a group (analogous to how a swarm of bees moves). To help picture how the group dynamics of bacteria operate, consider Figure 12.14. Here, a "capillary technique" for studying chemotaxis in populations of bacteria is shown. A capillary containing an attractant is placed in a medium with a bacterial suspension in Figure 12.14(a) and the bacteria then accumulate in the capillary containing the attractant as shown in Figure 12.14(b). Figure 12.14(c) shows what happens when the capillary contains neither attractant or repellant (i.e., a neutral environment relative to the medium it is placed in) and Figure 12.14(d) shows what happens when it contains a repellant.

Figure 12.14. Experiment showing how *E. coli* swarm towards nutrients, and away from noxious substances (figure taken from [21]).

It is interesting to note that *E. coli* and *S. typhimurium* can form intricate stable spatio-temporal patterns in certain semi-solid nutrient media. They can radially eat their way through a medium if placed together initially at its center. Moreover, under certain conditions they will secrete cell-to-cell attractant signals so that they will group and protect each other. These bacteria can "swarm." Other bacteria such as *M. Xanthus* exhibit relatively sophisticated social foraging behaviors involving several types of swarm behavior for protection, survival, and success in obtaining nutrients.

There are clear analogies with cooperative control of UAAVs:

■ Animals, organisms = UAAVs

■ Prey, nutrients = targets

■ Predators, noxious substances = threats

■ Environment = battlefield

Are these analogies useful? Biomimicry of social foraging of ants ([3]) has provided some concepts and algorithms useful for the solution of some engineering problems (where the key contribution is new algorithms for combinatorial optimization). What the relevance of the decision-making strategies of social foraging animals to the development of cooperative control strategies for UAAVs?

To begin to answer this question we have developed a model of the bacterial foraging process and have shown how a computer algorithm that emulates the behavior of a group of bacteria can solve a type of distributed optimization problem (in particular, not for combinatorial optimization, but one for a continuous cost function where there is no explicit analytical knowledge of the gradient). In particular, consider Figure 12.15 where we show the trajectories chosen by bacteria in a certain environment. Notice that they are successful in finding the areas where there are nutrients or prey (in our domain, targets) and in avoiding areas where there are noxious substances or predators (in our domain, threats). We have also simulated social foraging (swarming) for *E. coli* and *M. Xanthus* and studied mechanisms of aggregation and disintegration for these.

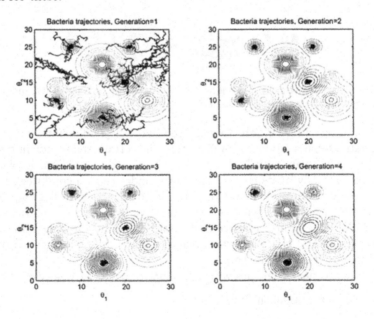

Figure 12.15.   Bacterial motion trajectories, generations 1-4, on contour plot.

6.   Concluding remarks

Advances in distributed computing and wireless communications have enabled the design of distributed agent systems. One of the key issues for a successful and wide deployment of such systems is the design of cooperative decision making and control strategies. Traditionally, feedback control methods have focused mostly on the design and analysis of centralized, inner-loop techniques. Decision and control of distributed vehicle systems requires a framework that is based more on cooperation between vehicles, and outer-loop schemes. In addition to cooperation, issues such as coordination, communication delays and robustness in the presence of losing one or more of the vehicles are crucial. In this paper, we have presented a framework for a special type of problem, the cooperative search. The proposed framework consists of two main components: learning the environment and using that knowledge to make intelligent high-level decisions on where to go (path planning) and what do to. We have presented some ideas regarding the design of a cooperative planning algorithm based on a recursive q-step ahead planning procedure and an interleaved planning technique. These ideas were illustrated with simulation studies by comparing them to a restricted random search and a standard search pattern. Moreover, we studied the stability of vehicular swarms to try to understand what types of communications are needed to achieve cooperative search and engagement, and characteristics that affect swarm aggregation and disintegration. Finally, we explored the utility of using biomimicry of social foraging strategies to develop coordination strategies.

References

[1] Ablavsky, V. and Snorrason, M. (2000). Optimal search for a moving target: a geometric approach. In *AIAA Guidance, Navigation, and Control Conference and Exhibit*, Denver, CO.

[2] Benkoski, S., Monticino, M., and Weisinger, J. (1991). A survery of the search theory literature. *Naval Research Logistics*, 38:469–494.

[3] Bonabeau, E., Dorigo, M., and Theraulaz, G. (1999). *Swarm Intelligence: From Natural to Artificial Systems*. Oxford Univ. Press, NY.

[4] Cameron, S. (1994). Obstacle avoidance and path planning. *Industrial Robot*, 21:9–14.

[5] Choset, H. and Pignon, P. (1997). Coverage path planning: the boustrophedon cellular decomposition. In *International Conference on Field and Service Robotics*, Canberra, Australia.

[6] Conn, A., Scheinberg, K., and Toint, P. (1997). Recent progress in unconstrained nonlinear optimization without derivatives. *Mathematical Programming*, 79:397–414.

[7] Danskin, J. (1968). A helicopter versus submarines search game. *Operations Research*, 16:509–517.

[8] Dell, R. and Eagle, J. (1996). Using multiple searchers in constrained-path moving-targer search problems. *Naval Research Logistics*, 43:463–480.

[9] Dennis, J. and Torczon, V. (1991). Direct search methods on parallel machines. *SIAM Journal Optimization*, 1:448–474.

[10] Eagle, J. and Yee, J. (1990). An optimal branch-and-bound procedure for the constrained path moving target search problem. *Operations Research*, 38:11–114.

[11] Gillen, D. and Jacques, D. (2000). Cooperative behavior schemes for improving the effectiveness of autonomous wide area search munitions. In *Proceedings of Workshop on Cooperative Control and Optimization*, University of Florida, Gainesville.

[12] Goldsmith, S. and Robinett, R. (1998). Collective search by mobile robots using alpha-beta coordination. In Drogoul, A., Tambe, M., and Fukuda, T., editors, *Collective Robotics*, pages 136–146. Springer Verlag: Berlin.

[13] Hert, S., Tiwari, S., and Lumelsky, V. (1996). A terrain-covering algorithm for auv. *Autonomous Robots*, 3:91–119.

[14] Hohzaki, R. and Iida, K. (1995a). An optimal search plan for a moving target when a search path is given. *Mathematica Japonica*, 41:175–184.

[15] Hohzaki, R. and Iida, K. (1995b). Path constrained search problem with reward criterion. *Journal of the Operations Research Society of Japan*, 38:254–264.

[16] Hohzaki, R. and Iida, K. (2000). A search game when a search path is given. *European Journal of Operational Reasearch*, 124:114–124.

[17] Jacques, D. and Leblanc, R. (1998). Effectiveness analysis for wide area search munitions. In *Proceedings of the AIAA Missile Sciences Conference*, Monterey, CA.

[18] Khatib, O. (1985). Real-time obstacle avoidance for manipulators and mobile robots. In *International Conference on Robotics and Automation*, pages 500–505, St. Louis.

[19] Koopman, B. (1980). *Search and Screening: General principles with Historical Application*. Pergarnon, New York.

[20] Lucidi, S. and Sciandrone, M. (1997). On the global convergence of derivative free methods for unconstrained optimization. In *Technical Report*. Univ.di Roma.

[21] Madigan, M., Martinko, J., and Parker, J. (1997). *Biology of Microorganisms*. Prentice Hall, NJ, 8 edition.

[22] Nelder, J. and Mead, R. (1965). A simplex method for function minimization. *Computer Journal*, 7:308–313.

[23] Passino, K. and Burgess, K. (1998). *Stability Analysis of Discrete Event Systems*. John Wiley and Sons Pub., New York.

[24] Passino, K. M. (2001). Biomimicry of bacterial foraging for distributed optimization and control. *IEEE Control Systems Magazine (to appear)*.

[25] Richardson, H. (1987). Search theory. In *Center for Naval Analyses*. N00-014-83-C-0725.

[26] Snorrason, M. and Norris, J. (1999). Vision based obstacle detection and path planetary rovers. In *Unmanned Ground Vehicle Technology II*, Orlanso, FL.

[27] Spires, S. and Goldsmith, S. (1998). Exhaustive geographic search with mobile robots along space-filling curves. In Drogoul, A., Tambe, M., and Fukuda, T., editors, *Collective Robotics*, pages 1–12. Springer Verlag: Berlin.

[28] Stephens, D. and Krebs, J. (1986). *Foraging Theory*. Princeton Univ. Press, Princeton, NJ.

[29] Stewart, T. (1980). Experience with a branch-and-bound algorithm for constrained searcher motion. In Haley, K. and Stone, L., editors, *Search Theory and Applications*, pages 247–253. Plenum Press, New York.

[30] Stone, L. (1975). *Theory of Optimal Search*. Acadamic Press, New York.

[31] Stone, L. (1983). The process of search planning: Current approachs and the continuing problems. *Operational Research*, 31:207–233.

[32] Torczon, V. (1991). On the convergence of the multidirectional search algorithm. *SIAM Journal Optimization*, 1:123–145.

[33] Torczon, V. (1997). On the convergence of pattern search algorithms. *SIAM Journal Optimization*, 7:1–25.

[34] Tsitsiklis, J. and Bertsekas, D. (1989). *Parallel and Distributed Computation*. Prentice-Hall, Inc., Engelwood Cliffs, NJ.

[35] Washburn, A. (1980). Search-evasion game in a fixed region. *Operations Research*, 28:1290–1298.

[36] Weaver, S., Baird, L., and Polycarpou, M. (1998). An analytical framework for local feedforward networks. *IEEE Transactions on Neural Networks*, 9(3):473–482.

Chapter 13

OPTIMAL RISK PATH ALGORITHMS*

Michael Zabarankin, Stanislav Uryasev, Panos Pardalos
Center for Applied Optimization,
Dept. of Industrial and Systems Engineering, University of Florida
zabarank@ufl.edu, uryasev@ise.ufl.edu, pardalos@ufl.edu

Abstract   Analytical and discrete optimization approaches for routing an aircraft
in a threat environment have been developed. Using these approaches,
an aircraft's optimal risk trajectory with a constraint on the path length
can be efficiently calculated. The analytical approach based on calculus of variations reduces the original risk optimization problem to the
system of nonlinear differential equations. In the case of a single radar-
installation, the solution of such a system is expressed by the elliptic
sine. The discrete optimization approach reformulates the problem as
the Weight Constrained Shortest Path Problem (WCSPP) for a grid
undirected graph. The WCSPP is efficiently solved by the Modified
Label Setting Algorithm (MLSA). Both approaches have been tested
with several numerical examples. Discrete nonsmooth solutions with
high precision coincide with exact continuous solutions. For the same
graph, time in which the discrete optimization algorithm computes the
optimal trajectory is independent of the number of radars. The discrete
approach is also efficient for solving the problem using different risk
functions.

Keywords: optimal risk path, length constraint, system of nonlinear differential
equations, analytical solution, discrete optimization approach, network
flow algorithms

Introduction

Optimal trajectory generation is a fundamental requirement for military aircraft flight management systems. These systems are required

*Research is supported by the Air Force grant F-08630-00-1-0001.

R. Murphey and P.M. Pardalos (eds.), Cooperative Control and Optimization, 273–298.
© 2002 *Kluwer Academic Publishers.*

to take advantage of all available information in order to perform integrated task processing and reduce pilot workloads. The systems should provide updates at regular time intervals sufficient for threat avoidance. To optimize flight trajectory in a threat environment, a model for the risk of aircraft detection is developed based on idealizing assumptions with respect to geometrical and physical aircraft properties. An ideal flight trajectory for military operations meets the mission requirements within the constraints of aircraft limitations while minimizing risk exposure. Several levels of information are used in the selection of such a flight path and velocity profile. The trajectory can be a function of mission requirements and the threat environment. The most challenging and general problem is finding a minimal risk path, which depends on the locations of radar-installations or Surface Air Missiles (SAM), and is subject to technological constraints such as limits on flying time, fuel capacity and trajectory length.

Optimal risk path generation is closely related to optimal search path planning for calculating a route for a searcher to maximize the probability of target detecting. Basic results of the optimal search theory can be found in [16, 17, 19]. The Special Issue on Search Theory [5] contains an introductory paper by Stone and Washburn describing state-of-the-art research in this area. This issue includes a survey of the literature on the search theory containing 239 references. The most frequently used setup is to partition a search space into rectangular cells and to allocate search efforts to these cells [16, 19, 22] (see, also, various approaches with this setup in [1, 10, 20]).

In spite of numerous studies in this area, only a few considered risk optimization problems with technological constraints. The main purpose of this paper is to develop fast algorithms for optimal risk path generation with a path length constraint. These algorithms are intended for solving the optimization problem in online applications. We have considered two approaches for solving the risk optimization problem: 1) analytical solution based on calculus of variations, and 2) discrete optimization.

Using calculus of variations apparatus, we have derived the system of nonlinear differential equations for finding the optimal risk trajectory with a path length constraint in a general case. We have obtained the analytical solution of this system of differential equations in the case of a single radar-installation. The solution is expressed by the elliptic sine. Using standard mathematical software, it is illustrated with several numerical examples. Although we have made significant progress in the development of the analytical approach, construction of an analytical solution in the case with an arbitrary number of radars is still an open issue.

We have developed the discrete optimization approach for optimal path generation with a path length constraint in the case with an arbitrary number of radars. Efficiency of discrete optimization approaches for calculating optimal risk path essentially depends upon risk definition, type of technological constraints, and the aircraft's trajectory approximation (see, for instance, [21] for discussions of these issues). Many of the previous studies on trajectory generation for military aircraft are concentrated on feasible direction algorithms and dynamic programming [6]. These methods tend to be computationally intense and, therefore, are not well suited for onboard applications. To improve computation time, John and Moore [21] used simple analytical risk functions. Based on such an approach, they developed lateral and vertical algorithms to optimize flight trajectory with respect to time, fuel, aircraft final position, and risk exposure. Nevertheless, these algorithms are not intended for solving optimization problems with technological constraints, such as a constraint on the path length. In this paper, we use simple analytical functions for defining the risk and reduce optimal risk path generation with a constraint on the length to a network flow problem. An admissible domain for a flight route is approximated by a grid undirected graph and an aircraft's trajectory is presented by a path in the graph. The optimal risk path problem is reformulated as the Weight Constrained Shortest Path Problem (WCSPP) for the grid undirected graph. Several network flow optimization algorithms are available for solving the WCSPP [7]. For our purposes, we use the Modified Label Setting Algorithm (MLSA) with a preprocessing procedure developed and implemented in C++ code by Boland and Dumitrescu [8]. A reader interested in the WCSPP and MLSA can find relevant information in [7, 8]. The original code solves problems with integer costs and weights. We have adopted and applied this code for solving risk optimization problems with real costs and weights. The efficiency of the discrete optimization approach is demonstrated with several numerical examples (various numbers of radars in different locations). Computation time in these examples is only several seconds, indicating that the MLSA is fast enough for use in online applications. However, the performance of the algorithm is evaluated only numerically. Theoretical evaluation of the algorithm's complexity is beyond the scope of this study. For the case with a single radar, we have compared analytical and numerical solutions and found that solutions coincide with high precision. This validates both analytical and discrete optimization approaches developed in this paper.

The paper is organized as follows: section 1 describes assumptions and the problem statement; section 2 presents the analytical approach; section 3 considers the discrete optimization approach; section 4 gives concluding remarks; the Appendix contains the derivation of the system of differential equations for finding an optimal risk path with a constraint on the path length in the case with an arbitrary number of radars and the derivation of the analytical solution for this system in the case with a single radar.

1. Model description and setup of the optimization problem

This paper develops a general methodology for optimal risk path planning with a constraint on the path length. However, to be more specific, we focus primarily on risks related to radar detection. For instance, for the case with two radars, Fig. 13.1 and 13.2 illustrate an unconstrained optimal trajectory and an optimal trajectory with a length constraint, respectively. Curves in these figures correspond to level sets of the risk with the maximal values in areas close to the radars. Risk declines when an aircraft is going away from radars and increases when the aircraft approaches them. The threat models used in this study do not refer to any specific military scenario. They are simple analytical models that characterize the general nature of threats.

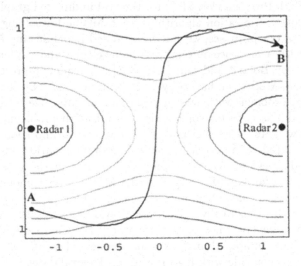

Figure 13.1. Aircraft flies from point A to point B trying to avoid two radars: optimal risk path without length constraint.

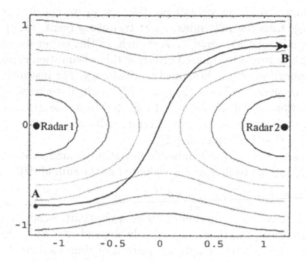

Figure 13.2. Aircraft flies from point A to point B trying to avoid two radars: optimal risk path with length constraint.

To formulate a risk path optimization problem the following assumptions are used:

1. Horizontal plane model. Aircraft position is considered to be in a horizontal plane only.

2. Radar detection of the aircraft does not depend on the aircraft's heading and climb angles.

3. Rotation angle does not depend on trajectory position.

4. An admissible domain for the aircraft's trajectory is assumed to be a detection area for all radar-installations. In other words, distance from the aircraft to each ith radar-installation is not greater than the ith radar maximum detection range, $R_{\max,i}$.

5. Risk is quantified in terms of the risk index per unit length for any particular aircraft location. The simplified threat model assumes that risk index r is proportional to risk factor σ and reciprocal to the squared distance from the aircraft position to the radar location (see, for instance, [21]). The risk factor, σ, depends on the radar's technical characteristics such as the maximum detection range, the minimum detectable signal, the transmitting power of the antenna, the antenna gain, and the wavelength of radar energy. It is considered that all of the radar's technical characteristics remain constant, hence, under such an assumption the risk factor, σ_i, is the constant for the ith radar. If $d_i = \sqrt{(x - a_i)^2 + (y - b_i)^2}$

is the distance from the aircraft position, (x, y), to the ith radar location, (a_i, b_i), then the risk index, r_i, at trajectory point, (x, y), is given by the formula $r_i(x, y) = \sigma_i d_i^{-2}$. Although, in this paper we considered that risk is reciprocal to squared distance, this assumption is not critical for the application of the developed methodology. For instance, the risk index can be expressed as $r_i(x, y) = \sigma_i d_i^{-4}$, which corresponds to the radar detection model with a signal reflected from the aircraft. In this case, the risk factor σ_i defines the aircraft radar cross section (RCS) for the ith radar (see, for instance, [18]).

6. At every point of the admissible deviation domain, the cumulative risk from N radar-installations is evaluated as the sum of risks from each ith radar, i. e. $r = \sum_{i=1}^{N} r_i = \sum_{i=1}^{N} \sigma_i d_i^{-2}$.

7. Aircraft velocity is assumed to be a constant, hence, time increment dt and unit length ds are linearly dependent: $ds = V_0 dt$. For convenience, we use unit length, ds, instead of time increment, dt.

8. For any particular aircraft location, (x, y), the risk per unit length, ds, is calculated as the product of the risk index and unit length, ds, i. e. $r\, ds = \sum_{i=1}^{N} \sigma_i d_i^{-2}\, ds$.

9. The risk R accumulated along a path P is presented by the expression $R(\mathrm{P}) = \int_{\mathrm{P}} r\, ds$.

Based on model assumptions $1 - 9$, the optimization problem with a constraint on the path length is formulated in the following way. Let N be the number of radars and (a_i, b_i) the location of the ith radar, where $i = \overline{1, N}$. The aircraft's departure and destination points are $A(x_1, y_1)$ and $B(x_2, y_2)$, respectively. The path P from A to B is associated with the integrated risk, $R(\mathrm{P})$, and the total length, $l(\mathrm{P})$. The optimal path P_* is such a path P, which minimizes $R(\mathrm{P})$ subject to length constraint $l(\mathrm{P}) \leq l_*$. The optimization problem is presented in the form

$$\min_{\mathrm{P}} R(\mathrm{P}) \qquad\qquad (1)$$
$$s.\,t. \quad l(\mathrm{P}) \leq l_*,$$

where $R(\mathrm{P})$ and $l(\mathrm{P})$ are defined by the expressions

$$R(\mathrm{P}) = \int_{A}^{B} \sum_{i=1}^{N} \sigma_i d_i^{-2}\, ds, \qquad\qquad (2)$$

$$l(\mathrm{P}) = \int\limits_A^B ds. \tag{3}$$

To solve problem (1)-(3), analytical and discrete optimization approaches are considered.

2. Analytical solution approach for the risk path optimization problem

We use calculus of variations apparatus to obtain an analytical solution for the risk optimization problem with a constraint on the path length. This approach reduces the original problem to the system of differential equations with respect to coordinates of the optimal trajectory. The complexity of this system depends upon the type of constraints and the manner in which the trajectory is defined.

The constraint on the trajectory length given in integral form (3) leads to an isoperimetric problem of calculus of variations (general theory is given in [11, 12]). We prefer not to use integral constraints in the formulation of the problem. To reduce constraint (3) to an algebraic form, coordinates of an unknown curve are presented as functions of the current length, s, of the curve, i. e. $x = x(s)$, $y = y(s)$. Such a parameterization is also known as the natural definition of a curve. The relation between differentials of the curve arc length, ds, and curve coordinates, dx and dy, such that $ds^2 = dx^2 + dy^2$, is reduced to an additional nonholonomic constraint for derivatives of $x(s)$ and $y(s)$. In other words, if we denote $\dot{x}(s) = \frac{d}{ds}x(s)$ and $\dot{y}(s) = \frac{d}{ds}y(s)$, then $\dot{x}(s)$ and $\dot{y}(s)$ should satisfy the equality $(\dot{x}(s))^2 + (\dot{y}(s))^2 = 1$. The distance between the ith radar location, (a_i, b_i), and the aircraft position, $(x(s), y(s))$, is defined by the expression $d_i = \sqrt{(x(s) - a_i)^2 + (y(s) - b_i)^2}$. With calculus of variations, problem (1)-(3) is reduced to minimization of functional $R(l)$ with respect to unknown curve $(x(s), y(s))$, where $0 \leq s \leq l$, with boundary conditions (x_1, y_1), (x_2, y_2) and constraint on the trajectory length, $l \leq l_*$. Problem (1)-(3) is reformulated in the form

$$\min_{(x,y)} \int\limits_0^l \sum_{i=1}^N \sigma_i d_i^{-2}\, ds, \tag{4}$$

subject to boundary conditions

$$\begin{aligned} x(0) &= x_1, & x(l) &= x_2, \\ y(0) &= y_1, & y(l) &= y_2, \end{aligned} \tag{5}$$

and constraints

$$(\dot{x}(s))^2 + (\dot{y}(s))^2 = 1, \tag{6}$$

$$l \le l_*. \tag{7}$$

Constraint $l \le l_*$ is different from $l = l_*$, which is considered in classical calculus of variations (see, for instance, [11, 12]). In the case when $l \le l_*$, it is a problem of calculus of variations with a moveable end point. The total variation of functional (4) must include the variation of the path length, l, and, therefore, value l is determined from an additional condition.

Let us denote the risk function in integral (4) by function L, which depends on variables x and y,

$$L\left(x(s), y(s)\right) = \sum_{i=1}^{N} \sigma_i d_i^{-2}. \tag{8}$$

Calculus of variations problem (4)-(7) for finding the optimal trajectory, $(x(s), y(s))$, is reduced to the system of differential equations

$$\begin{cases} L'_x - \frac{d}{ds}(\dot{x}L) = \lambda_L \ddot{x} \\ L'_y - \frac{d}{ds}(\dot{y}L) = \lambda_L \ddot{y} \end{cases} \tag{9}$$

with boundary conditions (5) and constraint (7), where λ_L is a nonnegative constant appearing due to length constraint (7). A detailed derivation of system (9) and the analytical solution for this system in the case of a single radar are included in the Appendix.

Since equation (6) is the first integral of system (9), constraint (6) and system (9) are not independent. While construction of an analytical solution of system (9) for a general form of function L is an open issue, the numerical solution for system (9) with boundary conditions (5), satisfying $l = l_*$, can be efficiently calculated by the method of weighted residuals [2, 15]. However, a numerical solution of system (9) is beyond the scope of this paper.

We have derived the analytical solution of system (9) in the case of a single radar. Without loss of generality, it is assumed that the radar is located at the origin of the system of coordinates, i. e. $a_1 = 0$, $b_1 = 0$, and that the risk factor, σ, equals 1. To simplify calculations, the polar system of coordinates is used. The polar radius, ρ, and polar angle, Ψ, are related to Cartesian coordinates x and y in the following way

$$x(s) = \rho(s) \cos \Psi(s),$$
$$y(s) = \rho(s) \sin \Psi(s).$$

Function L is rewritten as

$$L\left(\rho(s), \Psi(s)\right) = \rho^{-2}. \tag{10}$$

The optimal path has bounded length even in the case when length constraint (7) is relaxed. This is caused by the fact that function L does not converge rapidly enough to zero when ρ tends to infinity. The optimal solution for optimization problem (4)-(6) without constraint on the path length is presented by an arc of the circle

$$\left(x - (2a)^{-1} \sin C\right)^2 + \left(y - (2a)^{-1} \cos C\right)^2 = (2a)^{-2}, \tag{11}$$

where constants a and C are determined from boundary conditions (5). In the case when the constraint on the length is relaxed, the radar location, (a_1, b_1), and the aircraft's departure and destination points, (x_1, y_1) and (x_2, y_2), lie on the same circle. Denoting $\rho_1 = \sqrt{x_1^2 + y_1^2}$, $\rho_2 = \sqrt{x_2^2 + y_2^2}$ and the angle between vectors (x_1, y_1), (x_2, y_2) by ϑ, such that

$$\vartheta = \arccos\left(\frac{x_1 x_2 + y_1 y_2}{\rho_1 \rho_2}\right), \tag{12}$$

the optimal risk, R_*, and unconstrained path length, $\bar{l}$, are expressed in terms of ρ_1, ρ_2 and ϑ

$$R_* = \frac{1}{\rho_1 \rho_2} \sqrt{(x_2 - x_1)^2 + (y_2 - y_1)^2}, \tag{13}$$

$$\bar{l} = \frac{\vartheta}{\sin \vartheta} \sqrt{(x_2 - x_1)^2 + (y_2 - y_1)^2}. \tag{14}$$

If $\bar{l} \leq l_*$ then length constraint (7) is inactive and l coincides with $\bar{l}$. The optimal solution is determined by (11)-(14). In the case when $l_* \leq \bar{l}$, constraint (7) becomes active and l coincides with l_*. Using definition of the elliptic sine (see [3])

$$\operatorname{sn}[u, \kappa] = \sin\left(am(u, \kappa)\right) = \sin \phi, \qquad u = \int_0^\phi \frac{dt}{\sqrt{1 - \kappa^2 \sin^2 t}},$$

the optimal solution for the case $l_* \leq \bar{l}$, i. e. $l = l_*$, is presented in the form

$$\rho(\Psi) = \tfrac{1+\kappa}{a} \operatorname{sn}\left[\tfrac{1}{1+\kappa}\Psi + C, \kappa\right],$$

or

$$x(\Psi) = \tfrac{1+\kappa}{a} \, \text{sn} \left[\tfrac{1}{1+\kappa}\Psi + C, \; \kappa\right] \cos \Psi,$$
$$y(\Psi) = \tfrac{1+\kappa}{a} \, \text{sn} \left[\tfrac{1}{1+\kappa}\Psi + C, \; \kappa\right] \sin \Psi, \tag{15}$$

with boundary conditions

$$x^2 + y^2 = \rho_1^2, \quad \Psi = \Psi_1 = \arccos \tfrac{x_1}{\rho_1},$$
$$x^2 + y^2 = \rho_2^2, \quad \Psi = \Psi_2 = \arccos \tfrac{x_2}{\rho_2}, \tag{16}$$

where values ρ_1, ρ_2 are defined above and $0 \leq \Psi \leq \pi$.

According to (15), the optimal risk, R, and optimal path length, l, are given

$$R = \tfrac{a}{(1+\kappa)^2} \int_{\Psi_1}^{\Psi_2} \left(\kappa + \text{sn}^{-2}\left[\tfrac{1}{1+\kappa}\Psi + C, \; \kappa\right]\right) d\Psi, \tag{17}$$

$$l = a^{-1} \int_{\Psi_1}^{\Psi_2} \left(1 + \kappa \, \text{sn}^2\left[\tfrac{1}{1+\kappa}\Psi + C, \; \kappa\right]\right) d\Psi. \tag{18}$$

Constants a, κ, and C are determined from boundary conditions (16) and equality (18) under condition $l = l_*$.

First, the value of $\bar{l}$ should be calculated; if $\bar{l} \leq l_*$, the optimal solution is given by (11)-(14), and if $l_* \leq \bar{l}$ the optimal path is determined by (15)-(18).

To illustrate the analytical solution for the case of a single radar, we consider the following numerical example. Suppose that the aircraft's departure and destination points are A and B with coordinates $(x_1, y_1) =$ $(-0.25, 0.25)$, $(x_2, y_2) = (1.75, 0.25)$, respectively, and suppose that the maximum value for the trajectory length is $l_* = 3.2$. The radar is located at $(0,0)$ and the risk factor $\sigma = 1$. Values ϑ and $\bar{l}$ computed according to (12), (14) are $\vartheta = \tfrac{\pi}{4} + \arctan 7$ and $\bar{l} = 5.536$. The length constraint is active, since $\bar{l} \geq l_*$. Parameters a, κ and, C can be calculated using standard software; we have used package *MATHEMATICA 4*. For this case, $\kappa = 0.7378$, $a = 0.9414$, $C = 3.1419$. In Fig. 13.3, curve 1 (circle arc) represents the optimal path without constraint on the length. The risk for the path and the path length are equal to 3.2 and 5.536, respectively. Curve 2 is the optimal path with constraint on the length, $l_* = 3.2$. The risk for this path is 3.326. We compare the optimal risks and trajectory lengths not only for these two curves, but also for the straight line between points A and B. The length of line AB (curve 3),

l_{AB}, equals 2.0, while the risk accumulated along this line is equal to 8.857.

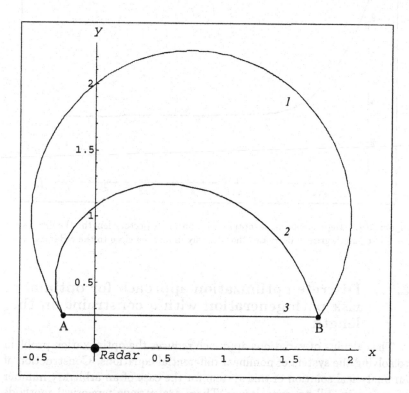

Figure 13.3. Optimal trajectories obtained by the analytical approach: curve 1 — optimal path without length constraint, $R= 3.2$, $l= 5.536$; curve 2 — optimal path with length constraint, $R= 3.326$, $l= 3.2$; curve 3 — shortest path, $R= 8.857$, $l= 2.0$

Function $R_{min} = R(l_*)$, $l_{AB} \leq l_* \leq \bar{l}$, is concave and bounded, see Fig. 13.4.

An obvious conclusion from this particular example is that the increase of the curve length considerably affects the risk only in an area close to the radar location.

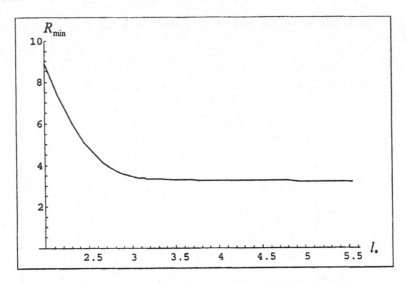

Figure 13.4. Dependence of the optimal risk on the trajectory length: the increase of the curve length greatly decreases the risk only in an area close to the radar location.

3. Discrete optimization approach for optimal risk path generation with a constraint on the length

The calculus of variations approach reduces the optimization problem to solving the system of nonlinear differential equations. Construction of an analytical solution of this system for the case of an arbitrary number of radars is still an open issue. There are various numerical methods approximating a solution of the system, but that is not the focus of this paper. We propose the discrete optimization approach to directly solve the original problem. This approach reduces optimal risk path generation with a constraint on the length to the Weight Constrained Shortest Path Problem for a grid undirected graph. In the case of an arbitrary number of radars, the WCSPP can be efficiently solved by network flow optimization algorithms. However, the computation time of these algorithms exponentially depends upon the precision prespecified for the optimal trajectory.

We assume the admissible deviation domain for the aircraft's trajectory to be an undirected graph $G = (V, A)$, where $V = \{1, \ldots, n\}$ is the set consisting of n nodes and A is the set of undirected arcs. A trajectory $(x(.), y(.))$ is approximated by a path P in the graph G, where path P is defined as a sequence of nodes $\langle j_0, j_1, \ldots, j_p \rangle$ such that $j_0 = A$, $j_p = B$

and $\langle j_{k-1}, j_k \rangle \in A$ for all k from 1 to p. To formulate (1)-(3) as a network optimization problem we use discrete approximation for formulas (2) and (3) determining the risk and trajectory length, respectively,

$$\int_A^B \left(\sum_{i=1}^N \sigma_i \, d_i^{-2} \right) ds = \sum_{k=1}^p \sum_{i=1}^N \sigma_i \, r_{i, \, j_{k-1} \, j_k} \Delta s_{j_{k-1} \, j_k}, \qquad (19)$$

$$\int_A^B ds = \sum_{k=1}^p \Delta s_{j_{k-1} \, j_k}, \qquad (20)$$

where $\Delta s_{j_{k-1} \, j_k}$ is the length of arc $\langle j_{k-1}, j_k \rangle$ and $r_{i, \, j_{k-1} \, j_k}$ denotes the risk index for the arc $\langle j_{k-1}, j_k \rangle$. If $x(j_k)$ and $y(j_k)$ are x and y coordinates for node j_k then arc length $\Delta s_{j_{k-1} \, j_k}$ is defined by the expression

$$\Delta s_{j_{k-1} \, j_k} = \sqrt{(x(j_k) - x(j_{k-1}))^2 + (y(j_k) - y(j_{k-1}))^2}. \qquad (21)$$

To derive the formula for the risk index $r_{i, \, j_{k-1} \, j_k}$ we compute the risk accumulated along the arc $\langle j_{k-1}, j_k \rangle$ from the radar located at (a_i, b_i) (the risk factor σ_i is omitted for convenience)

$$\int_{j_{k-1}}^{j_k} d_i^{-2} \, ds = \int_{x(j_{k-1})}^{x(j_k)} \frac{\sqrt{1 + (y_x')^2}}{(x - a_i)^2 + (y - b_i)^2} \, dx.$$

Since $y = y(x)$ is the straight line determined by the points $(x(j_{k-1}), y(j_{k-1}))$ and $(x(j_k), y(j_k))$, the integral above is rewritten as

$$\int_{j_{k-1}}^{j_k} d_i^{-2} \, ds = \frac{\left(\arctan \left(\frac{y(j_k) - b_i}{x(j_k) - a_i} \right) - \arctan \left(\frac{y(j_{k-1}) - b_i}{x(j_{k-1}) - a_i} \right) \right) \Delta s_{j_{k-1} \, j_k}}{(x(j_{k-1}) - a_i)(y(j_k) - b_i) - (x(j_k) - a_i)(y(j_{k-1}) - b_i)}.$$

The last formula has a simple interpretation. If we denote vector $(x(j_k) - a_i, y(j_k) - b_i)$ by $\mathbf{r}_{i, \, j_k}$, its length by $\|\mathbf{r}_{i, \, j_k}\|$ and the angle between two vectors $\mathbf{r}_{i, \, j_{k-1}}$ and $\mathbf{r}_{i, \, j_k}$ by $\vartheta_{i, \, j_{k-1} j_k}$, where $\vartheta_{i, \, j_{k-1} j_k}$ satisfies condition $0 \leq \vartheta_{i, \, j_{k-1} j_k} \leq \pi$, then the last formula turns to the following form

$$\int_{j_{k-1}}^{j_k} d_i^{-2} \, ds = \frac{\vartheta_{i, \, j_{k-1} j_k}}{\sin \vartheta_{i, \, j_{k-1} j_k}} \frac{\Delta s_{j_{k-1} \, j_k}}{\|\mathbf{r}_{i, \, j_{k-1}}\| \cdot \|\mathbf{r}_{i, \, j_k}\|},$$

and the risk index $r_{i,\,j_{k-1}\,j_k}$ is determined as

$$r_{i,\,j_{k-1}\,j_k} = \frac{\vartheta_{i,\,j_{k-1}j_k}}{\sin\vartheta_{i,\,j_{k-1}j_k}}\|r_{i,\,j_{k-1}}\|^{-1}\cdot\|r_{i,\,j_k}\|^{-1}. \qquad (22)$$

When $\vartheta_{i,\,j_{k-1}j_k}$ tends to zero, the risk index has limit value $\|r_{i,\,j_{k-1}}\|^{-1}\cdot$ $\|r_{i,\,j_k}\|^{-1}$. In the limit case $\Delta s_{j_{k-1}\,j_k}\to 0$, i. e. $j_{k-1}\to j_k$, we have $\vartheta_{i,\,j_{k-1}j_k}\to 0$, $\|r_{i,\,j_{k-1}}\|\to\|r_{i,\,j_k}\|$ and expression (22) coincides with the risk index definition for point j_k.

Fig. 13.5 illustrates a network flow example for solving the risk minimization problem. The thick broken line is a path in the area with two radars and $\langle j_{k-1},j_k\rangle$ is an arc of this path. The distance between nodes j_{k-1} and j_k is arc length $\Delta s_{j_{k-1}\,j_k}$. Values $\|r_{1,\,j_{k-1}}\|$, $\|r_{1,\,j_k}\|$ define distances from radar 1 to nodes j_{k-1} and j_k, respectively. Magnitude $\vartheta_{1,\,j_{k-1}j_k}$ corresponds to the angle between vectors $r_{1,\,j_{k-1}}$ and $r_{1,\,j_k}$.

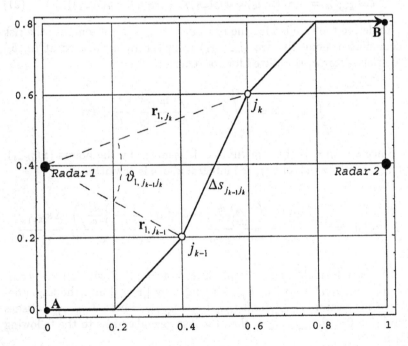

Figure 13.5. A network flow example for solving the risk minimization problem: the thick broken line AB is a path of the aircraft.

Using substitution

$$c_{j_{k-1} \, j_k} = \sum_{i=1}^{N} \sigma_i \, r_{i, \, j_{k-1} \, j_k} \Delta s_{j_{k-1} \, j_k}, \qquad (23)$$

values $R(\mathrm{P})$ and $l(\mathrm{P})$ are rearranged in the form

$$R(\mathrm{P}) = \sum_{k=1}^{p} c_{j_{k-1} \, j_k}, \qquad (24)$$

$$l(\mathrm{P}) = \sum_{k=1}^{p} \Delta s_{j_{k-1} \, j_k}. \qquad (25)$$

Thus, each arc $\langle j_{k-1}, j_k \rangle \in A$ is associated with its length $\Delta s_{j_{k-1} \, j_k}$ and nonnegative cost $c_{j_{k-1} \, j_k}$, defined by (21) and (23), respectively. Considering value $\Delta s_{j_{k-1} \, j_k}$ as the arc's weight we will use $R(\mathrm{P})$ to denote the cost of the path P and $l(\mathrm{P})$ to denote the total weight accumulated along that path. The path P is weight feasible if the total weight $l(\mathrm{P})$is at most l_*, i. e. $l(\mathrm{P}) \leq l_*$. The Weight Constrained Shortest Path Problem (WCSPP) is formulated in the following way. It is required to find such a feasible path P from point A to point B that minimizes cost $R(\mathrm{P})$

$$\min_{\mathrm{P}} \sum_{k=1}^{p} c_{j_{k-1} \, j_k}$$
$$s.\,t. \ \ \sum_{k=1}^{p} \Delta s_{j_{k-1} \, j_k} \leq l_*. \qquad (26)$$

The WCSPP (26) is closely related to the Shortest Path Problem with Time Windows (SPPTW) and also to the Resource Constrained Shortest Path Problem (RCSPP), which uses a vector of weights, or resources, rather than a scalar. These problems are solved in column generation approaches for Vehicle Routing Problems with Time Windows (VRPTW) and in long-haul aircraft routing problems. Algorithms for solving the WCSPP are divided into three major categories: label-setting algorithms based on dynamic programming methods, scaling algorithms, and algorithms based on the Lagrangean relaxation approach. The label setting algorithm is the most efficient in the case when the weights are positive [9]. The subgradient optimization [4] and cutting plane [13] methods are the core of the Lagrangean relaxation algorithm, which is efficient for solving the Lagrangean dual problem of the WCSPP in the case of one resource. Scaling algorithms use two fully polynomial approximation schemes for the WCSPP based on cost scaling and rounding [14]. The first scheme is a geometric bisection search whereas the second one iteratively extends paths. To solve the WCSPP, defined by (26), we use the Modified Label Setting Algorithm (MLSA) with a preprocessing procedure [8].

The Preprocessing Procedure and Label Setting Algorithm (LSA) are two consecutive stages of the MLSA. The LSA is the core of the MLSA, which integrates information obtained in preprocessing. The objective of the preprocessing procedure is to reduce the original graph by eliminating all arcs and nodes such that any path containing them is infeasible or does not improve current cost upper bound. To discuss the algorithm in detail, let us denote the arc's nodes j_{k-1} and j_k by i and j, respectively. For each node i, we consider the path obtained by appending the least cost path from the source node s to i to the least cost path from i to the sink node t. If the total cost accumulated along the new path is at least the current cost upper bound, then the use of node i cannot improve a known feasible solution. Hence, node i and all arcs incident to it can be deleted from the graph. If the total cost is less than the upper bound and the path is feasible, then the upper bound can be updated and the process continues with the improved upper bound. Similar, for each arc $\langle i, j \rangle$, we consider the path obtained by appending the least cost path from s to i to the least cost path from j to t, via arc $\langle i, j \rangle$. If the total cost accumulated along the new path is at least equal to the current cost upper bound, then we can delete arc $\langle i, j \rangle$ from the graph. If the total cost is less than the upper bound and the path is feasible then the upper bound can be updated. The preprocessing procedure is presented in the pseudo-code form below.

Preprocessing Algorithm for the WCSPP

Step 0: Let $U = C(n-1)$ where $C = \max\limits_{\langle i,j \rangle \in A} c_{ij}$.

Step 1: Find the shortest paths from source node $s = A$ with arc costs given by c_{ij}. Let Q_{sj}^c be the least cost path from s to j and α_j^c be the cost of the path: $\alpha_j^c = R(Q_{sj}^c)$.
If there is no path from s to the sink node $t = B$ **then** stop; the problem is infeasible.
If $(l(Q_{st}^c) \leq l_*)$ **then** Q_{st}^c is the optimal path.

Step 2: Find the shortest paths from all nodes to t with arc costs given by c_{ij}. Let Q_{jt}^c be the least cost path from j to t and β_j^c be the cost of the path: $\beta_j^c = R(Q_{jt}^c)$.

Step 3: Find the shortest paths from s to all nodes with arc lengths given by Δs_{ij}. Q_{sj}^l is the shortest path from s to j and α_j^l is the length of this path: $\alpha_j^l = l(Q_{sj}^l)$.

$\underline{\text{If}}$ $(l(Q_{st}^l) > l_*)$ $\underline{\text{then}}$ stop; the problem is infeasible.

$\underline{\text{If}}$ $(l(Q_{st}^l) \leq l_*)$ and $(R(Q_{st}^l) < U)$ $\underline{\text{then}}$ set $U = R(Q_{st}^l)$.

Step 4: Find the shortest paths from all nodes to t with the arc lengths given by Δs_{ij}. Q_{jt}^l is the least length path from j to t and β_j^l is the length of this path: $\beta_j^l = l(Q_{jt}^l)$.

Step 5: $\underline{\text{For}}$ all $j \in V \backslash \{s, t\}$ $\underline{\text{do}}$
$\underline{\text{if}}$ $(\alpha_j^l + \beta_j^l > l_*)$ $\underline{\text{then}}$ delete node j and all arcs incident to it;
$\underline{\text{if}}$ $(\alpha_j^c + \beta_j^c \geq U)$ $\underline{\text{then}}$ delete node j and all arcs incident to it;
$\underline{\text{end}}$

Step 6: $\underline{\text{For}}$ all $\langle i, j \rangle \in A$ $\underline{\text{do}}$
$\underline{\text{if}}$ $(\alpha_i^l + \Delta s_{ij} + \beta_j^l > l_*)$ $\underline{\text{then}}$ delete $\langle i, j \rangle$
$\underline{\text{else if}}$ $(\alpha_i^c + c_{ij} + \beta_j^c \geq U)$ $\underline{\text{then}}$ delete $\langle i, j \rangle$
$\underline{\text{else if}}$ $(l(Q_{si}^c) + \Delta s_{ij} + l(Q_{jt}^c) \leq l_*)$ $\underline{\text{then}}$ $U = \alpha_i^c + c_{ij} + \beta_j^c$;
$\underline{\text{end}}$

Step 7: $\underline{\text{If}}$ during steps 5 and 6 the graph changed $\underline{\text{then}}$ $\underline{\text{goto}}$ Step 1, $\underline{\text{else}}$ set $L = \alpha_t^c$ and stop.

End.

The next stage after the preprocessing procedure is the Label Setting Algorithm. The idea of the algorithm is to use a set of labels for each node and compare the labels to one another. Each label on a node represents a different path from node s to that node and consists of a pair of numbers representing the cost and weight of the corresponding path. No labels having the same cost are stored and for each label on a node, any other label on that node with a lower cost must have a greater weight. Let I_i be the index set of labels on node i and for each $k \in I_i$ let P_i^k denote a path from s to i with weight W_i^k and cost C_i^k. Pair (W_i^k, C_i^k) is the label of node i and P_i^k is the path corresponding to it. For two labels (W_i^k, C_i^k) and (W_i^q, C_i^q), corresponding to two different paths P_i^k and P_i^q, respectively, (W_i^k, C_i^k) dominates (W_i^q, C_i^q) if $W_i^k \leq W_i^q$, $C_i^k \leq C_i^q$, and the labels are not equal. Label (W_i^k, C_i^k) is efficient if it is not dominated by any other label at node i, i. e. if $(l(P), R(P))$ does not dominate (W_i^k, C_i^k) for all paths P from s to i. A path is efficient if the label it corresponds to is efficient. The LSA finds all efficient labels in every node. Starting without any labels on any node, except for label $(0, 0)$ on node s, the algorithm extends the set of all labels by treating an existing label on a node, that is, by extending the corresponding path along all outgoing arcs. Let L_i be the set of labels on node i and let $T_i \subseteq I_i$ index the labels on node i, which have been treated. The algorithm proceeds until all labels have been treated, i. e. until $I_i \backslash T_i = \emptyset$ for all $i \in V \backslash \{t\}$.

The Modified Label Setting Algorithm (MLSA)

Step 0: *Initialization*

Run Preprocessing Algorithm for the WCSPP to find U, β_j^c, β_j^l and $Q_{jt}^c \quad \forall j \in V \backslash \{t\}$.

Set $L_s = \{(0, 0)\}$ and $L_i = \emptyset$ for all $i \in V \backslash \{s\}$.

Initialize I_i accordingly for each $i \in V$.

Set $T_i = \emptyset$ for each $i \in V$.

Step 1: *Selection of the label to be treated*

If $\bigcup_{i \in V} (I_i \backslash T_i) = \emptyset$ **then** stop; all efficient labels have been generated.

Else choose $i \in V$ and $k \in I_i \backslash T_i$ so that W_i^k is minimal.

Step 2: *Treatment of label* (W_i^k, C_i^k)

For all $\langle i, j \rangle \in A$ **do**

    If $(W_i^k + \Delta s_{ij} + \beta_j^l \leq l_*)$

        If $(C_i^k + c_{ij} + \beta_j^c < U)$

            If $(W_i^k + \Delta s_{ij}, C_i^k + c_{ij})$ is not dominated
by $(W_j^q, C_j^q) \quad \forall q \in I_j$

            then set $L_j = L_j \cup \{(W_i^k + \Delta s_{ij}, C_i^k + c_{ij})\}$
and update I_j

            If $(W_i^k + \Delta s_{ij} + l(Q_{jt}^c) \leq l_*)$ **then** $U = C_i^k + c_{ij} + \beta_j^c$.

end

Step 3: Set $T_i = T_i \cup \{k\}$, **goto** to Step 1.

End.

The MLSA was implemented in C++ code by Boland and Dumitrescu [8]. The code was originally designed for solving the WCSPP with integer costs and integer weights. Under the assumption of cost and weight integrality, it was shown that the WCSPP is a NP-hard problem. In the case of real costs and weights, no specific evaluation for the complexity of the MLSA is provided. It is considered that in a worse case, the WCSPP can be solved in the time exponentially depending on the number of arcs. The analysis of the MLSA's complexity is beyond the scope of this paper.

The original code is adopted for solving the WCSPP with real costs and real weights assigned by (22) and (23), respectively.

For numerical examples, we use an undirected graph with nodes located in the cross-points of a square grid with the side length of T relative units. A set of arcs assigned for each node in the graph is shown in Fig. 13.6. This set consists of 32 arcs. Each arc is counted twice, since the graph is undirected. In Fig. 13.6, "1" represents four horizontal and four vertical arcs with the same length of Δs, "2" denotes eight diagonal arcs with the same length of $\sqrt{2}\,\Delta s$, and "3" denotes sixteen long diagonal arcs with the same length of $\sqrt{5}\,\Delta s$. If m is.the number of arcs on the grid side, i. e. $m = T/\Delta s$, then $(m+1)^2$ is the number of nodes and $4m\,(4m-1)$ is the number of arcs in the graph.

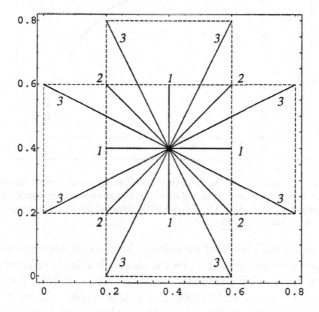

Figure 13.6. A set of arcs assigned for each node in the graph: 1 — horizontal and vertical arcs, 2 — diagonal arcs, 3 — long diagonal arcs.

To compare the analytical and discrete optimization approaches, we consider the previous example with a single radar. The radar location, the boundary value for the length constraint and the departure and destination points for the aircraft are exactly the same. The graph has the following characteristics: $T = 2.1$, $\Delta s = 0.05$, $m = 42$. There are 1849 nodes and 28056 arcs in this graph. All calculations have been performed using a PC 800 MHz with RAM of 256 Mb. The computation time is approximately 6 *sec*. Fig. 13.7 compares the analytical and discrete optimization solutions. The smooth curve is the optimal trajectory obtained by the analytical approach. The value of the risk accumulated along this trajectory equals 3.326. The nonsmooth curve is the optimal trajectory obtained by the discrete optimization approach. In this case, the trajectory has length 3.196 and risk value 3.360. For the discrete optimization solution, the relative error of the optimal risk value is about 1% . This validates both approaches.

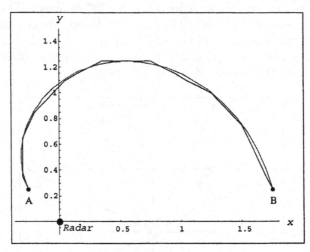

Figure 13.7. Comparison of the analytical and discrete optimization solutions: the smooth curve is the optimal path obtained by the analytical approach, $R=$ 3.326, $l=$ 3.2; the nonsmooth curve is the optimal path obtained by the discrete optimization approach, $R=$ 3.360, $l=$ 3.196

To estimate time required for calculating the optimal trajectory with higher precision, we again consider the case with a single radar using grid undirected graph with parameters: $T = 2.1$, $\Delta s = 0.025$, and $m = 84$. There are 7225 nodes and 112560 arcs in this graph. The solution time is about 200 *sec*. Fig. 13.8 compares the analytical solution and discrete optimization solutions with different precision (the discrete optimization solution with higher precision is depicted by a dashed curve). The dashed trajectory has length 3.199 and accumulated risk 3.357. The relative risk error for the dashed curve in comparison with the risk value for the analytical solution is about 0.9% . This indicates that the higher precision solution does not essentially improve the lower precision solution and that practically both trajectories are identical. In this case, the calculation time for the higher precision solution was about thirty times greater than for the lower precision solution. However, risk was improved by only 0.1% . This conclusion relates only to the considered example. In a general case, such an inference should be made based on the evaluation of the MLSA's complexity.

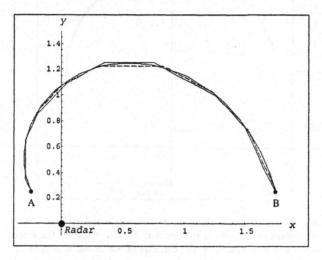

Figure 13.8. Comparison of the discrete optimization solutions with different precision: the dashed curve is the optimal discrete solution with higher precision, $R= 3.357$, $l= 3.199$; the solid nonsmooth curve is the optimal discrete solution with lower precision, $R= 3.360$, $l= 3.196$; the smooth curve is the analytical solution, $R= 3.326$, $l= 3.2$

Computation time of the MLSA does not depend on the number of radars; it depends only on the number of arcs. To demonstrate this advantage of the discrete optimization approach, we use the same graph for calculating the optimal risk trajectories with a different number of radars. In Fig. 13.9-13.12, the optimal risk trajectories are computed in the same time interval. In all considered examples, grid undirected graph with parameters $T = 2.0$, $\Delta s = 0.02$, $m = 100$ is used, and the risk factors σ_i are assumed to be equal to 1.

4.   Concluding remarks

We have developed analytical and discrete optimization approaches for calculating an optimal risk path with a constraint on the path length. We have studied optimization problems with the risk index in the form σd^{-2} (σ is the risk factor, d is the distance to the radar). However, the

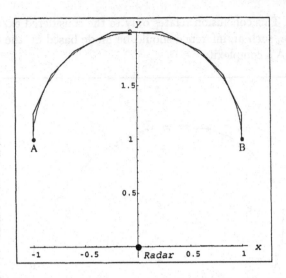

Figure 13.9. Optimal path in the case with a single radar; $T = 2.0$, $\Delta s = 0.02$, $m = 100$. The nonsmooth curve is the optimal discrete solution; the smooth curve is the analytical solution obtained by the calculus of variations approach.

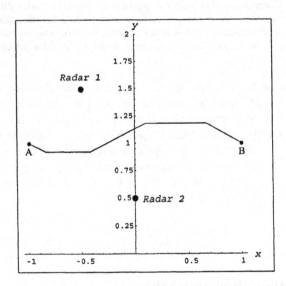

Figure 13.10. Optimal path obtained by discrete optimization in the case with two radars; $T = 2.0$, $\Delta s = 0.02$, $m = 100$.

developed methodology is quite general and it can be used to generate

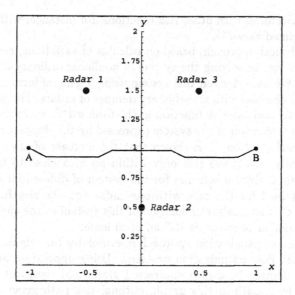

Figure 13.11. Optimal path, symmetric location of three radars; $T = 2.0$, $\Delta s = 0.02$, $m = 100$.

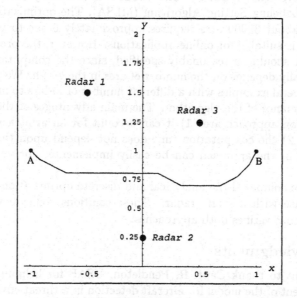

Figure 13.12. Optimal path, asymmetric location of three radars; $T = 2.0$, $\Delta s = 0.02$, $m = 100$.

optimal trajectories with other risk functions, for instance, with the risk function defined as σd^{-4}.

The analytical approach, based on calculus of variations, reduces the original problem to solving the system of nonlinear ordinary differential equations. We have derived this system using a general form of the risk function for the case with an arbitrary number of radars. For the case of a single radar and the risk function in the form σd^{-2}, we have obtained the analytical solution of the system expressed by the elliptic sine. Using the analytical solution, it is shown that the increase of the trajectory length greatly affects the risk only within an area close to the radar. Although an analytical solution for the system of differential equations can be obtained for the case with one radar and the risk function in the form σd^{-4}, an analytical solution of this system in the case with an arbitrary number of radars is still an open issue.

The discrete optimization approach reformulates the original problem as a network flow optimization problem. Using approximation for the admissible domain by a grid undirected graph and representation of a trajectory by a path in this graph, optimal risk path generation with a constraint on the length is reduced to the Weight Constrained Shortest Path Problem (WCSPP). The WCSPP is efficiently solved by the Modified Labeling Setting Algorithm (MLSA). The optimization problem with about 30000 arcs requires approximately 6 *sec* to be solved. This time is suitable for online applications. However, the precision for the MLSA should be reasonably specified, since the computation time exponentially depends on the number of arcs in the graph. We have considered several examples with a different number of radars to investigate the performance of the algorithm. The main advantages of the discrete optimization approach are: 1) it can account for an arbitrary number of radars; 2) the computation time does not depend upon the number of radars; 3) the approach can be easily implemented for various risk functions.

We have compared the analytical and discrete optimization solutions for the case with a single radar. These solutions coincide with high precision that verifies both approaches.

Acknowledgments

We want to thank Capt. R. Pendleton, USAF for helping with the development of the model for aircraft detection in a threat environment. We are also grateful to Prof. N. Boland and I. Dumitrescu for the informative discussions and for providing C++ code for the Modified Labeling Setting Algorithm that was used for conducting numerical experiments.

References

[1] Assaf, D. and Sharlin-Bilitzky, A. (1994). *Dynamic Search for a Moving Target.* Journal for Applied Probability, Vol. 31, No. 2, pp. 438 – 457.

[2] Aziz, A. K. (1975). *Numerical Solution of Boundary Value Problems for Ordinary Differential Equations.* New York: Academic Press.

[3] Bateman, H. and Erdelyi, A. (1955). *Higher Transcendental functions.* Vol. 3, Mc Graw-Hill Book Company, Inc.

[4] Beasley, J.E. and Christofides, N. (1989). *An Algorithm for the Resource Constrained Shortest Path Problem.* Networks 19, pp. 379 – 394.

[5] Benkoski, S.J., Monticino, M.G. and Weisinger J.R. (1991). *A Survey of the Search Theory Literature.* Naval Research Logistics, Vol. 38, No. 4, pp. 468 – 494.

[6] Chan, Y.K. and Foddy, M. (1985). *Real Time Optimal Flight Path Generation by Storage of Massive Data Bases.* Proceedings of the IEEE NEACON 1985, Institute of Electrical and Electronics Engineers, New York, pp. 516 – 521.

[7] Dumitrescu, I. and Boland, N. (2001). *Algorithms for the Weight Constrained Shortest Path Problem.* ITOR, Vol. 8, pp. 15 – 29.

[8] Dumitrescu, I. and Boland, N. (2001). *Improving Preprocessing, Labelling and Scaling Algorithms for the Weight Constrained Shortest Path Problem.* Submitted for publication to Networks.

[9] Desrochers, M. and Soumis, F. (1988). *A Generalized Permanent Labeling Algorithm for the Shortest Path Problem with Time Windows.* INFOR 26, pp. 191 – 212.

[10] Eagle, J.N. and Yee, J.R. (1990). *An Optimal Branch-and-Bound Procedure for the Constrained Path, Moving Target Search Problem.* Operations research, Vol. 38, No. 1, pp. 110 – 114.

[11] Elsgohs, L.E. (1961). *Calculus of Variations.* Franklin Book Company, Inc.

[12] Gelfand, I.M., Richard A. Silverman, Fomin, S.V. (2000). *Calculus of Variations.* Dover Publications, Inc.

[13] Handler, G.Y. and Zang, I. (1980). *A Dual Algorithm for the Constrained Shortest Path Problem.* Networks 10, pp. 293 – 309.

[14] Hassin, R. (1992). *Approximated Schemes for the Restricted Shortest Path Problem.* Mathematics of Operations Research 17, pp. 36 – 42.

[15] Ince, E.L. and Sneddon, I.N. (1987). *The Solution of Ordinary Differential Equations.* Halsted Press.

[16] Koopman, B.O. (1980). *Search and Screening: general principles with historical applications.* NY: Elmsford, Pergamon Press.

[17] Mangel, M. (1984). *Search Theory.* Lecture Notes. Berlin: Springer-Verlag.

[18] Skolnik, M.I. (1990). *Radar Handbook,* 2nd ed. New York: McGraw-Hill Book Company, Inc.

[19] Stone, L.D. (1975). *Theory of Optimal Search.* New York, San Francisco, London: Academic Press.

[20] Thomas, L.C. and Eagle, J.N. (1995). *Criteria and Approximate Methods for Path-Constrained Moving-Target Search Problems.* Naval Research Logistics, Vol. 42, pp. 27 – 38.

[21] Vian, J.L. and More, J.R. (1989). *Trajectory Optimization with Risk Minimization for Military Aircraft.* AIAA, J. of Guidance, Control and Dynamics, Vol. 12, No. 3, pp. 311 – 317.

[22] Washburn, A.R. (1983). *Search for a Moving Target: The FAB Algorithm.* Operations Research, Vol. 31, pp. 739 – 751.

Appendix

This appendix contains the derivation of the system of differential equations for determining an optimal risk path with a constraint on the path length and the analytical solution of this system for the case with a single radar.

We start with formulation of the calculus of variation problem with a nonholonomic constraint and a moveable end point

$$\min_{x,\,y} \ \Phi(x, y, \dot{x}, \dot{y}, l), \tag{13.A.1}$$

$$\Phi(x, y, \dot{x}, \dot{y}, l) = \int_0^l L\left(x(s), y(s), \dot{x}(s), \dot{y}(s)\right) ds, \tag{13.A.2}$$

$$\begin{aligned} x(0) &= x_1, & x(l) &= x_2, \\ y(0) &= y_1, & y(l) &= y_2, \end{aligned} \tag{13.A.3}$$

$$\varphi\left(\dot{x}(s), \dot{y}(s)\right) = 0, \tag{13.A.4}$$

$$l \le l_*. \tag{13.A.5}$$

A necessary condition for the existence of a functional extremum requires the total variation of the functional to be equal to zero. Constraint (13.A.5) implies that this is the problem with the movable end point, $(x(l), y(l))$ (variation of the total curve length, l, is not equal to zero). It should be taken into account that variations δx and δy are dependent due to nonholonomic constraint (13.A.4). Usually, to separate differential expressions in the functional variation, the Lagrange multiplier method is used. However, compared to the traditional approach, here, the multiplier $\lambda(s)$ depends upon variable s, since two degrees of freedom, variables x and y, are used to formulate the problem. Applying the Lagrange multiplier method to problem (13.A.1)-(13.A.5), the total variation of functional (13.A.2) with constraints (13.A.4) and (13.A.5) is rearranged in the form

$$\begin{aligned} \delta\Phi &= \int_0^l \left(\left[L'_x - \tfrac{d}{ds} L'_{\dot{x}} - \tfrac{d}{ds}\left(\lambda(s)\,\varphi'_{\dot{x}}\right) \right] \delta x + \left[L'_y - \tfrac{d}{ds} L'_{\dot{y}} - \tfrac{d}{ds}\left(\lambda(s)\,\varphi'_{\dot{y}}\right) \right] \delta y \right) ds \\ &\quad + \left. \left(L - \dot{x} L'_{\dot{x}} - \dot{y} L'_{\dot{y}} - \lambda(s)\left[\dot{x}\varphi'_{\dot{x}} + \dot{y}\varphi'_{\dot{y}}\right] \right) \right|_{s=l} \delta l \equiv 0. \end{aligned} \tag{13.A.6}$$

Assuming variation δx to be independent, and choosing $\lambda(s)$ to turn the expression

$$L'_y - \tfrac{d}{ds} L'_{\dot{y}} - \tfrac{d}{ds}\left(\lambda(s)\,\varphi'_{\dot{y}}\right)$$

at the variation δy to zero, equality (13.A.6) is reduced to the system of differential equations

$$\begin{aligned} L'_x - \tfrac{d}{ds} L'_{\dot{x}} - \tfrac{d}{ds}\left(\lambda(s)\,\varphi'_{\dot{x}}\right) &= 0, \\ L'_y - \tfrac{d}{ds} L'_{\dot{y}} - \tfrac{d}{ds}\left(\lambda(s)\,\varphi'_{\dot{y}}\right) &= 0. \end{aligned} \tag{13.A.7}$$

The equation defining the moveable end point, l, is given by

$$\left. \left(L - \dot{x} L'_{\dot{x}} - \dot{y} L'_{\dot{y}} - \lambda(s)\left[\dot{x}\varphi'_{\dot{x}} + \dot{y}\varphi'_{\dot{y}}\right] \right) \right|_{s=l} = 0. \tag{13.A.8}$$

It can be shown that equations of system (13.A.7) have the first integral. Summing the first equation multiplied by $\dot{x}$ with the second one multiplied by $\dot{y}$, we have

$$\dot{x}\left(L'_x - \tfrac{d}{ds}L'_{\dot{x}} - \tfrac{d}{ds}\left(\lambda(s)\,\varphi'_{\dot{x}}\right)\right) + \dot{y}\left(L'_y - \tfrac{d}{ds}L'_{\dot{y}} - \tfrac{d}{ds}\left(\lambda(s)\,\varphi'_{\dot{y}}\right)\right) = 0.$$

The left-hand side of this equality is a total differential. After integration it turns to the expression

$$L - \dot{x}L'_{\dot{x}} - \dot{y}L'_{\dot{y}} - \lambda(s)\left(\dot{x}\varphi'_{\dot{x}} + \dot{y}\varphi'_{\dot{y}}\right) = const. \qquad (13.A.9)$$

Lagrange multiplier, $\lambda(s)$, is derived from (13.A.9)

$$\lambda(s) = \frac{L + \lambda_L - \dot{x}L'_{\dot{x}} - \dot{y}L'_{\dot{y}}}{\dot{x}\varphi'_{\dot{x}} + \dot{y}\varphi'_{\dot{y}}}, \qquad \lambda_L = -const \geq 0. \qquad (13.A.10)$$

Substituting (13.A.10) into (13.A.7), we obtain the system of differential equations for determining $x(s)$ and $y(s)$. In the case when constraint $l \leq l_*$ is active, i. e. $l = l_*$, equation (13.A.8) is excluded from the system for determining an optimal solution, since the total curve length is fixed and, therefore, the variation of l equals zero by definition. If constraint $l \leq l_*$ is inactive, then from (13.A.8) and (13.A.9) we have $\lambda_L = 0$.

For the case of optimization problem (1)-(3), function L defines the risk index at the point (x, y), and hence, it depends on variables x and y only,

$$L\left(x, y, \dot{x}, \dot{y}\right) \equiv L\left(x(s), y(s)\right) = \sum_{i=1}^{N} \frac{\sigma_i}{\left(x(s) - a_i\right)^2 + \left(y(s) - b_i\right)^2}. \qquad (13.A.11)$$

Function φ represents reformulated nonholonomic constraint (6)

$$\varphi\left(\dot{x}(s), \dot{y}(s)\right) = \left(\dot{x}(s)\right)^2 + \left(\dot{y}(s)\right)^2 - 1 = 0. \qquad (13.A.12)$$

Substituting value $\dot{x}\varphi'_{\dot{x}} + \dot{y}\varphi'_{\dot{y}} = 2$ into formula (13.A.10), Lagrange multiplier, $\lambda(s)$, is rewritten as

$$2\lambda(s) = L\left(x(s), y(s)\right) + \lambda_L. \qquad (13.A.13)$$

Using (13.A.7), (13.A.8), (13.A.11) and (13.A.13), the original optimization problem (1)-(3) for defining the optimal trajectory $(x(s), y(s))$ is reduced to the system of differential equations

$$\begin{cases} L'_x - \tfrac{d}{ds}\left(\dot{x}L\right) = \lambda_L \ddot{x} \\ L'_y - \tfrac{d}{ds}\left(\dot{y}L\right) = \lambda_L \ddot{y} \end{cases} \qquad (13.A.14)$$

with boundary conditions (13.A.3) and (13.A.5). Only two equations from (13.A.12), (13.A.14) are independent, since equation (13.A.12) is the first integral of system (13.A.14). A pair of independent equations is chosen from (13.A.12) and (13.A.14) to simplify the derivation of the solution. Certainly, the obtained solution must satisfy (13.A.12) and (13.A.14). Obtaining an analytical solution for system (13.A.14) in the case of an arbitrary number of radars is still an open issue. We present the analytical solution of system (13.A.14) in the case with a single radar. Without loss of generality, let us assume that the radar is located at the origin of the system of coordinates, i. e. $a_1 = 0$, $b_1 = 0$, and that the risk factor $\sigma = 1$. In such a case,

$$L(x, y) = d^{-2} = \left(x^2 + y^2\right)^{-1}. \qquad (13.A.15)$$

To simplify further transformations, we use the polar system of coordinates. The polar radius ρ and polar angle Ψ are related to Cartesian coordinates x and y in the following way

$$x(s) = \rho(s)\cos\Psi(s),$$
$$y(s) = \rho(s)\sin\Psi(s),$$
(13.A.16)

$$\dot{x} = \dot{\rho}\cos\Psi - \rho\dot{\Psi}\sin\Psi,$$
$$\dot{y} = \dot{\rho}\sin\Psi + \rho\dot{\Psi}\cos\Psi.$$
(13.A.17)

Using (13.A.16), (13.A.17) and rearranged function L

$$L(\rho(s), \Psi(s)) = \rho^{-2},$$

system (13.A.14) and equation (13.A.12) are converted to equalities

$$2\rho^{-3}\cos\Psi + \frac{d}{ds}\left((\dot{\rho}\cos\Psi - \rho\dot{\Psi}\sin\Psi)(\rho^{-2} + \lambda_L)\right) = 0,$$
$$2\rho^{-3}\sin\Psi + \frac{d}{ds}\left((\dot{\rho}\sin\Psi + \rho\dot{\Psi}\cos\Psi)(\rho^{-2} + \lambda_L)\right) = 0,$$
(13.A.18)

$$(\dot{\rho})^2 + \rho^2(\dot{\Psi})^2 = 1.$$
(13.A.19)

Equations of system (13.A.18) have the first integral. Subtracting the second equation multiplied by $\cos\Psi$ from the first one multiplied by $\sin\Psi$, we obtain

$$\sin\Psi\frac{d}{ds}\left((\dot{\rho}\cos\Psi - \rho\dot{\Psi}\sin\Psi)(\rho^{-2} + \lambda_L)\right)$$
$$- \cos\Psi\frac{d}{ds}\left((\dot{\rho}\sin\Psi + \rho\dot{\Psi}\cos\Psi)(\rho^{-2} + \lambda_L)\right) = 0.$$

After algebraic transformations

$$\frac{\ddot{\Psi}}{\dot{\Psi}} + 2\frac{\dot{\rho}}{\rho} + \frac{\frac{d}{ds}(\rho^{-2} + \lambda_L)}{(\rho^{-2} + \lambda_L)} = 0.$$

Being a total differential, the left-hand side of the equation above is integrated and the right-hand side is converted to an unknown constant a. Using the equality obtained by integration, value $\dot{\Psi}$ is presented as the function of ρ

$$\dot{\Psi} = \frac{a}{1 + \lambda_L \rho^2}, \qquad a > 0.$$
(13.A.20)

By substituting (13.A.20) into (13.A.18) and (13.A.19), we obtain the differential system for defining $\rho(s)$ and $\Psi(s)$

$$\begin{cases} (\dot{\rho})^2 + \rho^2(\dot{\Psi})^2 = 1 \\ \dot{\Psi} = \frac{a}{1 + \lambda_L \rho^2} \end{cases} \qquad a > 0.$$
(13.A.21)

To solve system (13.A.21), the relation between differentials ρ and Ψ is used, i. e. if $\rho = \rho(\Psi)$ then

$$\dot{\rho} = \frac{d\rho}{d\Psi}\dot{\Psi}.$$
(13.A.22)

Based on (13.A.21) and (13.A.22), function $\rho(\Psi)$ must satisfy the nonlinear ordinary differential equation

$$\left(\frac{d\rho}{d\Psi}\right)^2 = a^{-2}\left(1 + \lambda_L \rho^2\right)^2 - \rho^2.$$
(13.A.23)

The solution of (13.A.23) is reduced to the integral representation for $\Psi = \Psi(\rho)$, i. e.

$$\Psi = a \int \frac{d\rho}{\sqrt{\left(1 + \lambda_L\, \rho^2\right)^2 - a^2 \rho^2}} + C'. \qquad (13.A.24)$$

Taking into account condition $a^2 > 2\lambda_L$, we make the following change of the variable in integral (13.A.24)

$$\rho = \alpha\tau, \qquad \lambda_L^2 \rho^4 - \left(a^2 - 2\lambda_L\right)\rho^2 + 1 = \left(1 - \tau^2\right)\left(1 - \kappa^2\tau^2\right),$$

where α and κ are new constants defined through a and λ_L as

$$\kappa = \frac{a^2 - 2\lambda_L - a\sqrt{a^2 - 4\lambda_L}}{2\lambda_L}, \qquad \alpha = \sqrt{\frac{\kappa}{\lambda_L}}, \qquad 0 \le \kappa \le 1, \ \lambda_L \ge 0. \quad (13.A.25)$$

Function Ψ is expressed by the elliptic integral of the first kind

$$\Psi = a \int \frac{\alpha\, d\tau}{\sqrt{1 - \tau^2}\sqrt{1 - \kappa^2\tau^2}} + C' = a\sqrt{\frac{\kappa}{\lambda_L}}\, F\left[\arcsin\sqrt{\frac{\lambda_L}{\kappa}}\rho, \ \kappa\right] + C''. \quad (13.A.26)$$

Inverting function (13.A.26) with respect to variable ρ, and using the expression for the elliptic sine (see, for instance, [3]),

$$\mathrm{sn}\,[u, \kappa] = \sin\left(am(u, \kappa)\right) = \sin\phi, \qquad u = \int_0^\phi \frac{dt}{\sqrt{1 - \kappa^2 \sin^2 t}},$$

we obtain the solution for equation (13.A.23) in the form

$$\rho(\Psi) = \sqrt{\frac{\kappa}{\lambda_L}}\ \mathrm{sn}\left[\sqrt{\frac{\lambda_L}{a^2\kappa}}\Psi + C, \ \kappa\right], \qquad C = -\sqrt{\frac{\lambda_L}{a^2\kappa}}\, C''. \quad (13.A.27)$$

To simplify relations between constants λ_L, a, and κ in (13.A.25), parameter λ_L is expressed by a and κ

$$\lambda_L = \frac{a^2\kappa}{(1 + \kappa)^2}, \qquad 0 \le \kappa \le 1. \qquad (13.A.28)$$

Substituting the value of (13.A.28) into expression (13.A.27), the final representation for $\rho(\Psi)$ is

$$\rho(\Psi) = \frac{1+\kappa}{a}\ \mathrm{sn}\left[\frac{1}{1+\kappa}\Psi + C, \ \kappa\right]. \qquad (13.A.29)$$

Based on (13.A.16) and (13.A.29), we have the analytical solution for the optimal trajectory (x, y)

$$\begin{aligned} x(\Psi) &= \tfrac{1+\kappa}{a}\ \mathrm{sn}\left[\tfrac{1}{1+\kappa}\Psi + C, \ \kappa\right]\cos\Psi, \\ y(\Psi) &= \tfrac{1+\kappa}{a}\ \mathrm{sn}\left[\tfrac{1}{1+\kappa}\Psi + C, \ \kappa\right]\sin\Psi, \end{aligned} \qquad (13.A.30)$$

with boundary conditions

$$\begin{aligned} x^2 + y^2 &= \rho_1^2, \quad \Psi = \Psi_1 = \arccos\tfrac{x_1}{\rho_1}, \\ x^2 + y^2 &= \rho_2^2, \quad \Psi = \Psi_2 = \arccos\tfrac{x_2}{\rho_2}, \end{aligned} \qquad (13.A.31)$$

where $\rho_1 = \sqrt{x_1^2 + y_1^2}$, $\rho_2 = \sqrt{x_2^2 + y_2^2}$ and $0 \leq \Psi \leq \pi$ by definition.

Using the arc length differential in the form $ds = a^{-1} \left(1 + \lambda_L \, \rho^2\right) d\Psi$, derived from (13.A.20), the total length and optimal risk for the trajectory $(x(s), y(s))$ are

$$l = a^{-1} \int_{\Psi_1}^{\Psi_2} \left(1 + \kappa \, \mathrm{sn}^2 \left[\tfrac{1}{1+\kappa}\Psi + C, \; \kappa\right]\right) d\Psi, \qquad (13.A.32)$$

$$R = \tfrac{a}{(1+\kappa)^2} \int_{\Psi_1}^{\Psi_2} \left(\kappa + \mathrm{sn}^{-2}\left[\tfrac{1}{1+\kappa}\Psi + C, \; \kappa\right]\right) d\Psi. \qquad (13.A.33)$$

Values for a, κ, and C should be determined in terms of l_* and boundary conditions (x_1, y_1), (x_2, y_2). In the case when the length constraint is relaxed, the optimal path has bounded length denoted by $\bar{l}$. If $\bar{l} \leq l_*$, then the length constraint, $l \leq l_*$, is inactive and l coincides with $\bar{l}$. To determine the value of $\bar{l}$, the optimal path without constraint on the length should be calculated. For the case when the length constraint is relaxed we have $\lambda_L = 0$. This implies $\kappa = 0$ and the elliptic sine becomes $\mathrm{sn}\,[u, 0] = \sin u$. Expression (13.A.29) is simplified

$$\rho(\Psi) = a^{-1} \sin\left(\Psi + C\right),$$

and the optimal solution is presented

$$\begin{aligned} x(\Psi) &= (2a)^{-1} \left(\sin C + \sin\left(2\Psi + C\right)\right), \\ y(\Psi) &= (2a)^{-1} \left(\cos C - \cos\left(2\Psi + C\right)\right). \end{aligned} \qquad (13.A.34)$$

System (13.A.34) represents the circle in the parametric form. Excluding parameter Ψ from the system, we obtain the well-known representation for the circle

$$\left(x - (2a)^{-1} \sin C\right)^2 + \left(y - (2a)^{-1} \cos C\right)^2 = (2a)^{-2}. \qquad (13.A.35)$$

Based on solution (13.A.35), unknown constants a and C are determined from the boundary conditions (x_1, y_1) and (x_2, y_2). If ϑ is the angle between vectors (x_1, y_1) and (x_2, y_2)

$$\vartheta = \arccos\left(\frac{x_1 x_2 + y_1 y_2}{\sqrt{x_1^2 + y_1^2}\sqrt{x_2^2 + y_2^2}}\right), \qquad (13.A.36)$$

then

$$a = \frac{\sin \vartheta}{\sqrt{(x_2 - x_1)^2 + (y_2 - y_1)^2}}, \qquad (13.A.37)$$

and value $\bar{l}$ for the unconstrained path length is determined by a and ϑ

$$\bar{l} = \vartheta / a. \qquad (13.A.38)$$

Notice that a is always nonnegative, since formula (13.A.36) defines the value of ϑ in interval $0 \leq \vartheta \leq \pi$ and, therefore, $\sin \vartheta \geq 0$ for all ϑ.

Consequently, if $\bar{l} \leq l_*$ then the optimal solution is given by (13.A.35), (13.A.37), and in the case when $l_* \leq \bar{l}$, the optimal path is determined by system (13.A.30), where a, κ, and, C must satisfy equation (13.A.32) under condition $l = l_*$.

Applied Optimization

1. D.-Z. Du and D.F. Hsu (eds.): *Combinatorial Network Theory.* 1996
   ISBN 0-7923-3777-8
2. M.J. Panik: *Linear Programming: Mathematics, Theory and Algorithms.* 1996
   ISBN 0-7923-3782-4
3. R.B. Kearfott and V. Kreinovich (eds.): *Applications of Interval Computations.* 1996
   ISBN 0-7923-3847-2
4. N. Hritonenko and Y. Yatsenko: *Modeling and Optimimization of the Lifetime of Technology.* 1996
   ISBN 0-7923-4014-0
5. T. Terlaky (ed.): *Interior Point Methods of Mathematical Programming.* 1996
   ISBN 0-7923-4201-1
6. B. Jansen: *Interior Point Techniques in Optimization.* Complementarity, Sensitivity and Algorithms. 1997
   ISBN 0-7923-4430-8
7. A. Migdalas, P.M. Pardalos and S. Storøy (eds.): *Parallel Computing in Optimization.* 1997
   ISBN 0-7923-4583-5
8. F.A. Lootsma: *Fuzzy Logic for Planning and Decision Making.* 1997
   ISBN 0-7923-4681-5
9. J.A. dos Santos Gromicho: *Quasiconvex Optimization and Location Theory.* 1998
   ISBN 0-7923-4694-7
10. V. Kreinovich, A. Lakeyev, J. Rohn and P. Kahl: *Computational Complexity and Feasibility of Data Processing and Interval Computations.* 1998
    ISBN 0-7923-4865-6
11. J. Gil-Aluja: *The Interactive Management of Human Resources in Uncertainty.* 1998
    ISBN 0-7923-4886-9
12. C. Zopounidis and A.I. Dimitras: *Multicriteria Decision Aid Methods for the Prediction of Business Failure.* 1998
    ISBN 0-7923-4900-8
13. F. Giannessi, S. Komlósi and T. Rapcsák (eds.): *New Trends in Mathematical Programming.* Homage to Steven Vajda. 1998
    ISBN 0-7923-5036-7
14. Ya-xiang Yuan (ed.): *Advances in Nonlinear Programming.* Proceedings of the '96 International Conference on Nonlinear Programming. 1998
    ISBN 0-7923-5053-7
15. W.W. Hager and P.M. Pardalos: *Optimal Control.* Theory, Algorithms, and Applications. 1998
    ISBN 0-7923-5067-7
16. Gang Yu (ed.): *Industrial Applications of Combinatorial Optimization.* 1998
    ISBN 0-7923-5073-1
17. D. Braha and O. Maimon (eds.): *A Mathematical Theory of Design: Foundations, Algorithms and Applications.* 1998
    ISBN 0-7923-5079-0

Applied Optimization

Applied Optimization

Applied Optimization

KLUWER ACADEMIC PUBLISHERS – DORDRECHT / BOSTON / LONDON